CHINA'S
NEW ENERGY
REVOLUTION

**How The World Super Power is Fostering
Economic Development and Sustainable
GROWTH THROUGH THIN FILM
SOLAR TECHNOLOGY**

LI HEJUN

Chairman of Hanergy Holding Group

Foreword by Ai Feng
Former Editor-in-Chief of *Economic Daily*
Founder of Brandcn.com

New York   Chicago   San Francisco   Athens   London
Madrid   Mexico City   Milan   New Delhi
Singapore   Sydney   Toronto

Markono Print Media Pte Ltd

ISBN   978-0-07-183577-0
MHID   0-07-183577-6

e-ISBN   978-0-07-183525-1
e-MHID   0-07-183525-3

McGraw-Hill Education books are available at special quantity discounts to use as premiums and sales promotions or for use in corporate training programs. To contact a representative, please e-mail us at bulksales@mheducation.com.

CONTENTS

FOREWORD

I Believe in This Prediction

A prediction is made in this book that China will lead the world in the new energy revolution.

This is a bold prediction of great significance that is worth our close attention.

A prediction is not reality. When things become reality, people no longer need a prediction. The value of a prediction lies in its accurate and forward-looking quality. A prediction then becomes the direction and motivator of efforts.

Will the author's prediction materialize? This might be the book's biggest question for its readers, especially for readers from China.

My impression is that Li Hejun is good at predictions. His pet phrase is "I have a prediction." It is noteworthy that so far all of Li's previous predictions have been confirmed. When preparing to build the Jin'anqiao Hydropower Station, he predicted that the government would eventually allow private companies to enter the energy sector. When entering the PV industry, he predicted that thin-film technologies, which were not as popular as crystalline silicon at the time, would catch up quickly and therefore should be preferred. When the wave of incidents relating to the PV industry was yet to arrive, he had made the prediction that there would be a radical shake-up in the PV industry in China and in the world at large. All these predictions of his have been proved correct, one by one. They have also enabled Hanergy to become China's largest private clean energy company and the world's largest thin-film solar cell company in the first decade or so of the twenty-first century.

Today, the author is making an even bolder and overarching prediction: China will lead the world in this new energy revolution. Will this prediction materialize?

I have no doubt. This is not only because of the accuracy of the author's previous predictions. More importantly, this is an extremely convincing book.

The author discusses the new energy revolution at three levels.

First, at a global level, the fossil fuel–based energy system is nearing its end. Conditions have been created for new energy sources to take over. With both demand and preconditions, a new energy revolution is unstoppable.

Second, in China, where the biggest barrier to sustainability is an energy bottleneck formed in the fossil fuel era, new energy sources are gaining momentum to take over. With emerging advantages in accessibility, economic benefits, market demand, industry maturity, and policy incentives, it is entirely possible for China to lead the world in new energy development.

Third, at the industry level, the author delineates global trends in the PV industry while drawing lessons from abroad. By chronicling the history of PV technologies, the author demystifies the PV industry and identifies problems to be solved in the Chinese industry. He goes on to suggest strategies, approaches, and policies to facilitate the new energy revolution. The end result is a mini encyclopedia of PV industry strategies.

The author's reasoning is so factually accurate and logically powerful that you cannot help but agree with his prediction.

Of course, even an accurate prediction is only a possibility and needs our efforts to realize it. This is what the book leaves the reader with. I believe that as long as we join forces, the author's prediction will soon come true.

Ai Feng
Former Editor-in-Chief of *Economic Daily*
Founder of Brandcn.com

PREFACE

For China's photovoltaics (PV) industry, 2012 was both a troubled and a transformative year.

There was much bad news that year.

The world's production capacity for crystalline silicon solar cells has soared, but prices have plummeted. Behind this is China's huge new PV production capacity (it has quadrupled from 2009 to 2011) driven by a single-minded pursuit of quantity at the cost of creating vicious cycles.

After releasing results of its anti-dumping and anti-subsidies duty investigations on Chinese-made solar panels, the US has announced anti-dumping taxes on Chinese PV imports. The EU has also initiated its anti-dumping investigations against the Chinese PV industry. India has followed suit.

Against the backdrop of the global recession, as governments of European countries have lowered their subsidies for the PV industry, China's PV module manufacturers are facing significantly reduced revenues coupled with an end to loans from domestic banks. Several Chinese PV companies are now on the brink of bankruptcy.

Formerly one of the richest people in China, Shi Zhengrong, the one-time chairman and CEO of Suntech Power, has withdrawn into a secondary position. The founder of LDK Solar resigned as its CEO. Top executives of many other PV companies have similarly chosen to "back off." The PV industry has retreated into a state of fear and anxiety.

Despite the above, upon close observation, there was also much good news in 2012.

First Solar, a US company, achieved a net income of USD[1] 87.9 million in the third quarter of 2012 despite the continued loss-making circumstances in the crystalline silicon solar cell industry. Although this result was far short of the USD 196.5 million it made in the same period of 2011, it was still a good carry on of the USD 110.9 million it made in the second quarter of 2012.

In China, the *12th Five-Year Plan for the Solar Photovoltaic Industry* has been published. Measures have been implemented to enable distributed, grid-connected PV generation.

As China's largest private clean energy company, Hanergy has started to acquire PV companies overseas.

There is a life cycle to the prosperity or failure of any industry or business. A crisis is happening today in China's PV industry, which has been regarded as a booming industry ever since its earliest days. Is the doomsday scenario that some have pictured for the industry tenable? When a huge wave comes, some ride on it to get ahead whereas others are pushed backwards. If China's PV industry is facing a collective crisis, why are some companies making great strides ahead whereas others are quickly falling behind?

There is no absolute crisis nor absolute boom. If we dialectically think over the multiple positive and negative developments in China's PV industry in 2012, we may reach a drastically different conclusion.

Looking at the negative developments, China's PV industry was indeed beset with difficulties both at home and abroad.

In March 2012, the US Department of Commerce announced anti-subsidies duties on solar panel imports from China. It went on to announce anti-dumping duties on May 17. On November 7, the US International Trade Commission set final anti-dumping duties between 18.32% and 249.96% and final anti-subsidies duties between 14.78% and 15.97% on crystalline silicon PV cells and other imports from China. Europe has followed suit with applications to launch anti-dumping investigations against PV imports from China. On September 6, the E.U., a major destination for Chinese-made PV products, started investigations involving products worth 20 billion USD.

As 90% of China's polycrystalline silicon PV cell production is exported to overseas markets, the anti-dumping and anti-subsidies duty investigations by the US and the EU were deeply worrisome. None of the leading Chinese PV module manufacturers, including Yingli, Suntech Power, Trina Solar, and Canadian Solar, made a profit in 2012. What is worse is that many PV companies find themselves in the predicament of overstock, overcapacity and huge debts.

According to data from Maxim Group, a US investment management firm, in August 2012, China's top 10 PV companies accumulated a combined debt of USD 17.5 billion. The debt ratio of PV companies in general has exceeded a whopping 70%.

In the meantime, pessimism was pervasive in the mainstream media. The view that there is overcapacity in the PV industry quickly spread across the country after the annual sessions of the National People's Congress, creating two direct consequences.

The first consequence is that financial institutions suddenly stopped issuing loans for the PV industry. Once banks stopped issuing loans, some PV companies were pushed to the brink of bankruptcy. According to news reporting, Suntech Power, with a debt of USD 3.587 billion or an 81.8% debt ratio, saw its stock price plunge to USD 0.94 per share in the beginning of August 2012, compared to USD 80 per share in 2008. Due to inability to make payments, Suntech Power defaulted on convertible bonds in March 2013 and had to apply for bankruptcy.

The second consequence is that in the first quarter of 2013, Chinese-made PV solar cells basically quit the US market under the pressure of the investigations. PV companies were forced to seek differentiation: shifting away from manufacturing solar panels to making short-term investments in PV solar power station operations in Europe and the USA, in an attempt to sustain their value chain.

> In the past, we considered PV solar cells only a product for export and neglected its application within China in a new energy revolution. We have supported Europe's energy revolution by using huge amounts of energy here at home to produce PV products and then exporting them cheaply. The result was a staggering blow when the EU and the US launched trade investigations. Then, without understanding the difference between crystalline silicon cells and thin-film cells, we made a sweeping generalization about PV overcapacity, which has led to bank loan terminations. We have been hurt by others in our left hand, and then we have tied up our own right hand.

Looking at the positive developments, Chinese PV companies are also making great strides in developing second-generation PV technologies and acquiring PV companies overseas.

On June 5, 2012, Hanergy signed agreements with Q-Cells, a well-known solar energy company in Germany, to purchase the stock rights of Solibro, a subsidiary of Q-Cells. This was Hanergy's first move in its overseas M&A activities. Among the thin-film solar cells made by Solibro, a type of small cell has achieved the world's best CIGS conversion efficiency at 18.7%. Lars Stolt, considered the father of thin-film technologies, was the founder and CTO of Solibro. Today, he is a senior executive at Hanergy.

On January 9, 2013, Hanergy announced its acquisition of MiaSole, a US company that received investments of over USD 500 million from venture capitalists such as the legendary John Doerr in the past decade.

MiaSole was a typical example of an American PV company in Silicon Valley. The conversion efficiency of its mass-produced thin-film PV modules has reached 15.5%. The conversion efficiency of products under research and development has reached 17.6%, catching up with the conversion efficiency of crystalline silicon modules. Within two years, production cost is expected to decrease to USD 0.5 per watt.

Hanergy may well be able to realize John Doerr's dream of green development.

It is noteworthy that both Solibro and MiaSole had their own proprietary thin-film technology. As both are Hanergy companies now, they will be able to achieve synergy by sharing technologies without the former barriers of business confidentiality.

Reuters's analysis suggested that after this successful acquisition, Hanergy will be ready to compete with the world's largest thin-film PV cell manufacturer, First Solar. A penetrating analysis, indeed.

On July 25, 2013, Hanergy bought out Global Solar, another US company.

Hanergy's three acquisitions completed within a year have drawn wide attention from news media at home and abroad. Some foreign news outlets suggested that when the PV industry is at a low ebb, Chinese companies are taking the opportunity to pick up pearls; and that US PV startups are being saved from failure by larger Asian industrial companies.

Statistics speak louder than words: by the end of 2012, Hanergy owned seven thin-film technologies and had built nine thin-film production facilities; Hanergy's production capacity had reached 3GW,[2] surpassing First Solar's 2.8GW to become the world's largest thin-film solar energy company.

From the point of view of negative developments, a huge crisis seems to be looming, likely to negate all the technologies and growth models used by China's PV industry in the past. From the point of view of positive developments, the case of Hanergy is the best argument against the doomsday scenario some have pictured for China's PV industry. Instead, Chinese PV companies may have found the new technologies and growth models they have always needed to become innovative global leaders.

To gain a more accurate grasp of the PV industry in China and the world at large, we need a different and a broader perspective.

The Third Industrial Revolution by Jeremy Rifkin, an internationally renowned American futurist, provides us with a powerful set of coordinates.

In this book, Rifkin asserts that each and every industrial revolution in human history changes the world radically. Today, we are in the last stage of the second Industrial Revolution as well as the era of oil. The third Industrial Revolution has arrived, with new communication technologies converging with new energy regimes. Hundreds of millions of people will be able to produce their own green energy from their homes and offices and in factories, and share with others on an Internet of energy. Life and work will fundamentally change.

I was deeply moved by the author's insights. Reflecting upon my experience working in the new energy sector in the past 20 years and especially in the PV industry in recent years, I found Rifkin's theory very convincing.

A revolution in the energy sector is the most powerful propellant in each industrial revolution. In the second Industrial Revolution, at the core of its energy revolution was the replacement of coal by oil, though both are fossil fuels. In the third Industrial Revolution, new energy sources will replace fossil fuels. In particular, solar energy will play a critical role in the energy revolution this time.

Just as in the first two industrial revolutions, the third Industrial Revolution will bring about profound changes in how things are produced and how organizations are structured. As the basis of each country's competitiveness and the competitive landscape of global industries are being restructured, new big powers will emerge. China missed the first Industrial Revolution and was only able to catch the last train in the second Industrial Revolution. If it seizes the opportunities presented by the third Industrial Revolution, the realization of the Chinese Dream will no longer be far off.

Therefore, I have reaffirmed my faith in China's PV industry with a better understanding of how it can strategically serve the country.

From an industry standpoint, our choice is correct for a number of reasons. First, speaking of future trends in the energy sector, thin-film PV companies, as represented by Hanergy and First Solar, have chosen PV solar power among all the new energy sources as their focus. This means that we have grasped the direction of the energy revolution as the third Industrial Revolution arrives. Second, speaking of technology, thin-film PV companies, as represented by Hanergy and First Solar, are using the most promising technology they have mastered to promote the upgrade from first-generation PV (crystalline silicon) to second-generation PV (thin film) .

From a national strategy standpoint, visionaries in China have realized the strategic significance of the PV industry. Such a realization has led to a series of favorable industrial policies rolled out in 2012 and 2013.

May 23, 2012: At the State Council executive meeting, it was decided that steps would be taken to support the use of self-contained solar energy products in public spaces and households.

September 12, 2012: the *12th Five-Year Plan for Solar Power Development* was published. Installed capacity for PV power generation has been raised to 30–40 GW from 21GW.

September 14, 2012: The National Energy Administration issued the *Notice on Application of Large-scale Usage Demonstration Area of Distributed Solar PV Power Generation*.

October 26, 2012: The State Grid issued the *Guidelines on Providing Better Grid Connection Services for Distributed Photovoltaic Power Generation.*

November 9, 2012: The Ministry of Finance, the Ministry of Science and Technology, the Ministry of Housing and Urban–Rural Development, and the National Energy Administration jointly issued a notice about launching a second round of Golden Sun and Building–Integrated PV demonstration projects in 2012.

December 19, 2012: Then Premier Wen Jiabao held a State Council executive meeting to discuss policy measures for the healthy development of the PV industry.

July 15, 2013: The State Council issued *Several Opinions on Promoting the Healthy Development of the Photovoltaic Industry.*

July 31, 2013: The Ministry of Finance issued a notice on subsidies for distributed solar PV power generation based on the quantity of power generated.

......

How will negative factors such as the widespread panic about a doomsday scenario impact our PV industry? Does it mean that the industry still has a future but we need to readjust our strategies? Does it mean that we need a breakthrough in technology and growth pattern in the PV industry?

On the other hand, how will positive factors such as overseas M&A and favorable policy measures impact our PV industry? Does it mean that we have found a better growth pattern for the PV industry to become a world leader? Should we approach issues in the PV industry from a national strategy standpoint? Should the co-relation between the PV industry and national revival be incorporated as a part of the third Industrial Revolution theoretical framework?

These are the questions that I am going to discuss in this book. As a manager of a PV company, a practitioner in the new energy sector, and an eager contributor to our national revival, I am thinking out of the PV box. More importantly, I am seeking to contribute to the realization of the Chinese Dream.

The train of thoughts in this book is: What is the significance of a PV revolution to a new energy revolution? What is the significance of a new energy revolution to the third Industrial Revolution? What is the significance of the third Industrial Revolution to the rise of China? To summarize these questions: how will a new energy revolution with a PV revolution at its core facilitate the realization of the Chinese Dream?

I hope that we can embark on this thinking journey together with the following questions.

First, at a global level, our questions include:

- What are the fundamental changes brought about by the first two Industrial Revolutions?
- What do we think will be the major trend in the third Industrial Revolution?
- Is a new energy revolution at the core of the third Industrial Revolution? Why?
- Is a PV revolution at the core of the new energy revolution?
- Why will the PV revolution present an opportunity for China to lead the world in this industry?
- What are the US, the EU, Japan, and South Korea doing in this PV revolution?

Second, at a national level, our questions include:

- What are the advantages of China's PV industry?
- How should China position itself in the PV revolution?
- How can we facilitate the PV revolution using strategies, planning and policies?
- What changes will the PV revolution bring to China?

Third, at an industry level, our questions include:

- Does thin film stand for the future of the PV industry? Why?
- Is overcapacity an accurate generalization for the problems in China's PV industry?
- How can we increase domestic demand for PV power generation?
- How can we evaluate the pathway to grid parity in PV solar power?
- How can we build distributed power supply systems?

......

From different levels and standpoints, these questions form a tree. The trunk of the tree is major trends. On the branches, the destinies of companies, the development of the industry, and national strategies are interwoven and interconnected. One problem will lead to another and finally to wider consequences.

I would like to name this tree "PV Revolution." Despite the prevailing pessimism, I believe that we have found a new direction for the PV industry. The PV revolution will become the core and engine of the new energy revolution. By playing a critical role in the third Industrial Revolution, the PV revolution will deeply touch the way we live and work. In this process, China has every reason to get ahead by making the PV revolution an engine for national revival.

I firmly believe that in the future, solar power will replace coal, oil, and natural gas to become our main energy source. We will live under roofs installed with solar panels. We will enjoy fresh air and blue sky by using clean energy. We will enjoy a beautiful life.

This is my PV dream. It will illuminate the Chinese Dream.

HOLY FUCK GANGSTA

TRUTH

CHAPTER 1

NEW ENERGY REVOLUTION: CHINA'S CHOICE

Introduction

Why did the prosperity shown in the painting "Riverside Scene at Qingming Festival" by Zhang Zeduan (Song Dynasty) not carry over from Kaifeng, the capital city of the Song Dynasty, to Beijing, the capital of the Qing Dynasty, making the so-called "Celestial Empire" of Emperor Qianlong prosperous forever? Why could the electric light invented by Thomas Edison illuminate the whole world from America?

Humanity has experienced industrial revolutions twice and is now meeting the third one.

The prime power of each industrial revolution might seem to be technological inventions but actually they are energy revolutions. In the first industrial revolution, the British successfully substituted coal for firewood, realizing Britain's dream of becoming a great nation. In the second industrial revolution, the Americans substituted petroleum for coal and enjoyed the glory of leading the world.

What will the theme of energy revolution be in the third industrial revolution? The photovoltaic revolution has given us a possibility. The core competition mode in new energy, with solar power as its, representative, is different to that of traditional energies. It's not resource competition, but a competition in core technology. Whoever grasps the core technology will control energy.

The 21st century will witness China's rise. China's photovoltaic industry has been leading the world. Can China seize the opportunity in the energy revolution, and make the "Photovoltaic Dream" illuminate the "China Dream"?

Energy Substitution: The Core Power of Industrial Revolution

We've experienced industrial revolutions twice with magnificent upsurges in productivity. The driving force of the first one might seem to be the steam engine, but actually was coal; the driving force of the second one might seem to be electricity, but actually is petroleum. From firewood to coal to petroleum, the history of mankind has been changed by fossil fuels, while the future of mankind depends on renewable energy.

How will renewable energy replace fossil fuels? What's going to happen in the future?

1

Jeremy Rifkin's "Theory of the Five Pillars"

On May 24th, 2011, Jeremy Rifkin, who is an American futurist of international reputation and the author of the book *The Third Industrial Revolution*, attended the opening ceremony of the 50th Ministerial Conference of the OECD, where he proposed a crucial plan to government heads and ministers—the five pillar third Industrial Revolution economic development plan.

Over the past two decades, this farsighted professor at the Wharton School of The University of Pennsylvania has almost overturned the theoretical system and important ideas of his predecessors about the third Industrial Revolution.

Before Jeremy Rifkin, a universal definition of the third Industrial Revolution existed which was written into high school textbooks: the third revolution is another great stride for humankind in science and technology after the steam technology revolution (the first Industrial Revolution) and the electric technology revolution (the second Industrial Revolution) in the history of human civilization. It is a revolution of information control technology featuring inventions and their application in atomic energy, computing, space technology and bioengineering, and involving areas such as information technology, new energy technology, new material technology, biotechnology, space technology, and marine technology, among others.

Rifkin responds: "the concept in senior high school textbooks couldn't be more wrong." In his view, any Industrial Revolution includes two co-existing factors interacting with each other: an energy revolution and a revolution of information dissemination methods. Following this reasoning, the essence of the two earlier Industrial Revolutions is: during the first Industrial Revolution (1760s–1840s), communication technology underwent revolutionary changes from hand printing to steam engine–powered printing, with the latter capable of low cost large-scale printing and dissemination of information. This resembles the changes wrought by the Internet today: people use new communication systems to manage the new energy system based on coal; during the second Industrial Revolution (1870s–early 20th century), communication and energy joined hands again to integrate electricity, telephone and subsequently radio and television, so that more complex oil pipeline and highway networks could be managed to open the way for the emergence of urban culture.

In the 1990s, based on long research, Rifkin overturned his predecessors' understanding of the third Industrial Revolution, and proposed a new definition: the third revolution is a major social and economic transformation catalyzed by the combination of the Internet and renewable energy.

When economists studied the third Industrial Revolution in the past, some of them kept their eyes on energy, while others focused on communication. Rifkin combines the two creatively, and reaches a conclusion confidently: communication is the nervous system of the social organism, while energy is the blood. Nowadays, distributed information and communication technologies are making a powerful combination with distributed renewable energy, jointly giving birth to the real third Industrial Revolution.

To spread his "overwhelming" conclusion, Rifkin published a lot of papers in renowned international magazines such as *The Economist, The World Financial Review*, etc., and published a series of works such as *The End of Work, The Biotech Century, The Age of Access*, etc. Each book has been translated into more than 15 languages. *The Third Industrial Revolution* epitomized his theory.

Besides lecturing at Wharton and writing, Rifkin is also an activist. As of 2000, he's been shuttling across the Atlantic Ocean, lecturing, serving as consultant, organizing foundations and lobbying. Two-fifths of his time is spent in EU countries.

Rifkin has successfully made "the third Industrial Revolution" a term used by the political leaders of the EU. In 2006, he began to co-draft an economic development plan with senior officials of the European Parliament. In May 2007, the European Parliament released an official written statement, announcing a plan to take the third Industrial Revolution as a long-term economic focus and a road map for the EU's development. Currently, many institutions and member states of the Council of Europe are executing this road map.

At the opening ceremony in May 2011, Rifkin showed his latest research results—the Five-Pillar Economic Plan of the Third Industrial Revolution: (1) make the transition to renewable energy; (2) transform buildings in every continent into micro power plants to collect renewable energy; (3) apply hydrogen and other storage techniques in every building and the infrastructure to store intermittent energy; (4) make use of Internet technology to convert the power grid of every continent into an energy sharing network to regulate supply and demand for reasonable allocation; (5) turn transportation into plug-in and fuel cell vehicles whose power comes from the aforementioned power grid.

All the core vocabulary in the Five-Pillar Economic Plan is related to new energy—"renewable energy," "micro power plant," "storage technology," "energy sharing network," "fuel cell vehicle" ... i.e., the connection between the third Industrial Revolution and new energy is put on the agenda.

From Firewood to Coal to Petroleum

Let's follow Rifkin's theoretical framework to review the connection between the first two industrial revolutions and energy revolution.

In ancient times, humanity's main energy source was firewood. Although coal and petroleum were used sometimes, they were merely fuels for sporadic living and metal smelting, with very limited dynamic effect on human civilization. Thus, this age can be called "vegetal energy era."

The change in human society from the vegetal energy era to the fossil fuel era began in Britain in the 1860s. From the end of 16th century to the late 17th century, Britain's mining industry, particularly coal, had grown to a certain scale. Man (and animal) power could not meet the groundwater drainage demands in mining, while there was rich and cheap coal on site to be used as fuel. This reality pushed people to seek a new and stronger power source. In 1698, Thomas Savery from Devonshire invented the world's first steam engine. In 1712, Thomas Newcomen improved it and produced the "Newcomen steam engine." In 1769, Scottish inventor James Watt improved the technical design of the steam engine based on previous inventions, and produced the first modern steam engine.

Based on its predecessors' discourse, the significance of the "Watt steam engine" lies mainly in technological revolution— its cylinder, piston, flywheel, pendulum governor, valve and sealing element were all basic elements of a variety of modern machines. This series of technologies spawned the modern machine building industry. Then, modern thermodynamics emerged, laying a foundation for the steam turbine and internal combustion engine. It also pushed forward the development of the engineering industry, settled the critical problem of machine production, and generated huge progress in transportation.

In my view, the larger significance of inventing the steam engine is that it meant an end to the vegetal energy era, with firewood as the main power source, and the beginning of the fossil fuel era with coal dominant.

Steam engines pushed people to convert from manual labor to using powered machines on a large scale. The textile industry sprang up. The land, forest, and the food resources which the vegetal energy era relied upon couldn't support the development of large-scale industry and the demands of commercial transportation. With the development of the manufacturing and steel industries, mankind had to face the following problem: even if all the forest on earth were cut down, humanity's need for smelted iron ore still couldn't be met. Also, could a large enough source of energy be found to provide sufficient energy for steam engines?

Man had to turn to coal. In 1709, Abraham Darby successfully substituted cheap coke for charcoal that had begun to get scarce in Britain. The iron plant of Abraham's family had become the most successful iron-making and casting company in Britain in the 18th century. With the success of Darby, Coalbrookdale, a discarded ancient iron smelter, gradually developed into an iron and steel smelting center, and became a key industrial town finally.

In 1738, coal had been seen as the "Soul of the British Manufacturing Industry." The Anglo-Saxons, keen on navigation and business, started to lead the world in industrialization at that time.

In 1800, the coal and iron produced by Britain exceeded the production of all the other countries in the world. Coal production in Britain grew to 12 million tons in 1800 from 6 million tons in 1770, and then to 57 million tons in 1861. Likewise, iron production in Britain grew from 50,000 tons in 1770 to 130,000 tons in 1800, and then to 3.8 million tons in 1861. Iron had become cheap and plentiful enough to be used in general construction. Humanity had stepped into the steam age, the iron age, as well as the coal age.

This was the first industrial revolution in human history. From the perspective of energy, what had been done was the substitution of coal for firewood.

One big problem with fossil fuels is that the total amount available is reduced with exploitation; it is non-renewable. During the 100 years after 1820, countries taking coal as their main energy resource all face the reality of coal resources drying up.

In 1861, Britain produced 57 million tons of coal, and more than 100 million tons in 1865. In 1900, its coal production reached 225 million tons, a 13-fold increase from that in 1820 and almost 1.3 times that in 1865. On the eve of World War I, Britain's coal production reached a peak of 270 million tons, and then dropped to 240 million tons around 1929. Around 1950, Britain's coal production was just above 200 million tons. Around 2010, production was only about 20 million tons. However, its coal use had grown to nearly 20 *billion* tons, with the residual coal resources less than one-third of the total.

In the mid-19th century, Britain faced a severe "energy crisis" and "energy security" issue and the second Industrial Revolution went on the historical stage. This time, the key role in energy substitution was taken by petroleum.

The traditional view is that the prime content of the second substitution is the wide use of electricity: in 1831, British scientist Faraday discovered the electromagnetism induction phenomenon; in 1866, Siemens (Germany) invented the electric generator; in 1870, Gelam (Belgian) invented the electric motor. With rapid development of the electric power industry and appliance manufacture, mankind stepped into the "Electricity Age."

The traditional view holds that the second Industrial Revolution also included three elements: the creation of internal combustion engines and new transportation tools, the invention of means of communication, and the establishment of the chemical industry.

I believe that summing up the main thrust of "the second Industrial Revolution" with the four elements above is correct, but not penetrating enough. This kind of summary pays more attention to the technological revolution, but neglects the common driving factor hiding behind it—the energy revolution.

Let's start from the invention of the internal combustion engine.

In 1876, Nicolaus August Otto (Germany) made the first four-stroke internal combustion engine, taking coal gas as the fuel. In the mid-1880s, German inventor Karl Benz and his colleagues successfully developed the gasoline-powered internal combustion engine. In the 1890s, German engineer Rudolf Diesel designed an internal combustion engine with higher efficiency. It is called the diesel engine because it uses diesel as its fuel.

The invention of the internal combustion engine resolved the problem of what engine to use for transportation. In the late 19th century, a new type of transportation vehicle—the automobile—appeared. In the 1880s, Karl Benz successfully made the first automobile driven by a gasoline engine, and then established the world-famous Benz Motors Company. In 1896, Henry Ford (American) made his first four-wheeled automobile, and then set up the Ford Motor Company, creating the industrial production mode called Fordism.

Later on, the internal combustion locomotive, the ocean-going vessel, airplane and other machines that are driven by internal combustion engines quickly emerged. In 1903, the Wright brothers made the very first flight in an airplane, realizing humanity's dream of flying in the sky and symbolizing the dawn of a new transportation and communication age. Meanwhile, the invention of the internal combustion engine promoted the development of the oil industry, and the petrochemical industry emerged.

In short, electricity, with petroleum as its energy basis, became a new source of energy supplementing and replacing steam power. It is this change that brought mankind into the "Electricity Age." With petroleum gradually becoming the most basic fuel source; humanity opened a new transportation era. Likewise, the development of the petroleum industry catalyzed modern industry.

Compared with coal, petroleum has superior physical properties, generating twice as much heat as coal of the same quality and volume, and even three times as much in direct use. Petroleum is prone to vaporization and capable of continuous combustion. Thus, in the second Industrial Revolution, the volume of petroleum exploitation across the world grew bigger and bigger, and the status of petroleum increasingly prominent.

In terms of output, in 1870, global oil production was merely 800 000 tons, which soared to 20 million tons in 1900, 278 million tons in 1940, and 519 million tons in 1950.

In terms of the proportion between oil and coal, take America as an example. In 1900, America's oil production was 8.7 million tons, or about 6.3% of its coal production; in 1913, it reached 34 million tons, equivalent to 10.7% of the coal production; in 1920, it grew to 61 million tons, equivalent to 18.5% of its coal production; in 1950, it was 269 million tons, equivalent to 96% of its coal production.

Moreover, the oil age is also a period dominated by the US In 1920, oil production in the US accounted for about 10% of the world's total; in 1940, its oil production was 191 million tons, accounting for 68.7% of the world's total; and in 1950, the two figures reached 269 million tons and 51.8%, respectively.

Thus, the conclusion that I can draw boldly is: the core content of the second Industrial Revolution is actually an energy revolution, whose essence is that petroleum replaced coal as the dominant source of energy. The US led this transformation.

After looking back on the change over time from firewood to coal and then to petroleum, we can sum up the first two industrial revolutions from the angle of energy alternation. The 100-odd years after 1820 were the first phase of the fossil fuel age—the age of coal. During this phase, humanity established an economic system based on coal-dominant energy. In the early 20th century, the dominant position of coal began to be taken over by petroleum. Then humanity entered the second phase of the fossil fuel age—the age of petroleum, during which the economic system based on oil-dominant energy was set up. In other words, humanity's Industrial Revolution history is not only about the progression of technical economy, but also about energy substitution.

> Looking back in history, the first two Industrial Revolutions in human society all took energy substitution as their content and symbol. The third Industrial Revolution is no exception to this rule.

A New Substitution Is Happening Right Now

In the latter half of the 20th century, the concept of a new round of industrial revolution began to sprout in America. In the 1970s, American scholars started probing into "the third Industrial Revolution." Scholars such as Helfgott, Greenwood and Mowery noticed the influence of new technologies, information technology in particular, on workers' status in companies, industrial R & D structure, etc., and thus inferred that a new industrial revolution was around the corner.

With the coming of the 21st century, the depletion trend in petroleum and other fossil fuels grew increasingly prominent. The related global climate change poses a threat to human beings' continued existence. Meanwhile, the existing industrial economy model, driven by fossil fuels cannot support the world's continuing development. These things require us to seek a new model for human beings entering the "post-carbon" age. The theory of "the third Industrial Revolution" is formally put forward.

In the 21st century, there are two representative figures for the theory of the third Industrial Revolution. One is Jeremy Rifkin, the other is Paul Markillie.

Paul Markillie has paid close attention to the development of manufacturing technology and digital manufacturing for a long time. He thinks that the digital aspect of the third Industrial Revolution will bring a major change to the manufacturing model. Large-scale assembly-line manufacturing will be over. People will be able to follow their own will to design and produce things. The third Industrial Revolution might even bring in an anti-urbanization tide, in which a distributed and self-sufficient rural lifestyle replaces urban life.

As mentioned earlier, Rifkin's discourse and Paul Markillie's theory don't align. Markillie said that the Internet, green power and 3D printing technology are guiding capitalism into the third Industrial Revolution, with sustained and distributed development. The so-called third Industrial Revolution is that the combination of the Internet and renewable energy leads to a major shift in life, society and the economy.

In terms of "energy substitution," there are two fundamental differences between the third Industrial Revolution and the first two.

First, different content. The first two Industrial Revolutions substituted one fossil fuel for firewood and then one for another, while the third Industrial Revolution uses a new energy with a totally different nature (i.e., as a renewable energy) to substitute for fossil fuels. In the future, there will be no other energy to substitute for renewable energy. Therefore, the substitution this time can be called the "ultimate substitution."

Second, different purpose. Energy substitution in the third Industrial Revolution no longer makes the pursuit of wealth its ultimate goal. Instead, improving the quality of life and promoting the sustainable development of social civilization dominate.

Putting aside academic differences, we will look into the future of the third Industrial Revolution with the focus on renewable energy that is emphasized by Rifkin. Renewable energy is a concept closely related to new energy. The explanation of new energy by the United Nations Conference on New and Renewable Sources of Energy, held in 1980, is as follows:

Based on new technology and new materials, traditional renewable energy involves modern development and utilization, and substitutes inexhaustible renewable energy running in a loop for fossil fuels that are limited and cause pollution. The focus is on developing solar power, wind power, biomass energy, tidal energy, geothermal energy, hydrogen energy, and nuclear energy.

In brief, so-called new energy is non-conventional energy. It is a concept contrasting with traditional energy and conventional energy. Most of the time, this concept coincides with renewable energy, but it is also distinct sometimes. For example, hydro-energy is renewable, but because the development and utilization of hydropower has been around for a while, it is often seen as a traditional energy according to the current energy classification standard.

Fossil fuels are less and less adaptive to the demands of social progress. For example, energy shortages, power grid accidents, and soaring oil prices are common. Meanwhile, shortcomings include high carbon emissions, aggravating environmental pollution increasingly. Therefore, new energy, environmentally friendly and renewable, is receiving more and more attention from various countries.

On August 14, 2003, North America suffered the worst ever large-scale blackout, affecting one-quarter of the US. According to estimates by David Rosenberg, the chief economist of Merrill Lynch, the economy suffered a loss of about 25–30 billion dollars. This accident had a great impact on modern industrial civilization—even in a superpower where electric lights and the telephone were invented and the Empire State Building was lit up one hundred years ago, the security of the power grid is not stable.

In September 2004, the 19th World Energy Congress was held in Sydney. Twenty-five hundred government officials, industry leaders and scholars attended the conference, whose theme was "Delivering Sustainability: Opportunities and Challenges for the Energy Industry." At the conference, many experts agreed that soaring oil prices in the 21st century might be a long-term trend.

That is certainly the case. The peak period of the Age of Oil is 1950–1980, during which time oil was cheap, and supplies sufficient. But after 60-odd years of huge consumption, the global oil supply has entered a strategic drying-up period, with oil prices soaring.

Due to the skyrocketing of the price of oil, currently one-third of the world's population cannot enjoy modern energy services. Moreover, as demonstrated by the above examples, the current energy system, dominated by oil, can neither ensure energy security nor meet the requirement of environmental protection.

In January 2013, North and Central China suffered from severe smog continuously, particularly Beijing, where only 4 days out of that month saw good weather—the other days of that month being covered in dim haze. Fog and serious haze were caused by anthropogenic pollution, and the culprit is fossil-based energy consumption.

Table 1–1 Crude Oil Price Changes since 1970

Time	Price Change
1970	Saudi Arabia's official crude oil price was 1.8 dollars per barrel
1974 (the first oil crisis)	Crude oil price first broke through 10 dollars per barrel
1979 (the second oil crisis)	Crude oil price first broke through 20 dollars per barrel
1980	Crude oil price first broke through 30 dollars per barrel
Early 1981	International crude oil price reached 39 dollars per barrel
September 2004	Influenced by the Iraq war, international crude oil price broke through 40 dollars per barrel, and then continued to increase to 50 dollars per barrel
June 2005	International crude oil price first broke through 60 dollars per barrel
August 2005	Mexico suffered hurricane Katrina, international crude oil price first broke through 70 dollars per barrel
Sep. 12, 2007	International crude oil price first broke through 80 dollars per barrel, then the price continued to accelerate
Oct. 18, 2007	International crude oil price first broke through 90 dollars per barrel, and approached 100 dollars per barrel by year end
July 14, 2008	International crude oil price soared, crude oil futures at New York Mercantile Exchange reached a record high of 147.27 dollars per barrel
Jan. 21, 2009	International crude oil price fell back sharply due to the financial crisis. crude oil future's prices at New York Mercantile Exchange dropped to 33.20 dollars per barrel, a new low since April 2004
April 29, 2011	Crude oil futures at New York Mercantile Exchange rose up to 114.83 dollars per barrel

Data source: Sina Finance

It's not just China. Scientific reports around the world sounded a warning of an increasingly severe environmental crisis facing Earth. One radical view in the scientific community is that if the concentration of CO_2 cannot fall back from the 430 ppm recorded in 2008, partial melting of the Earth's ice cap will cause inundation. Unless human beings manage to reduce CO_2 emission to 5–10 billion tons in 2050, humanity will face calamity.

The energy crisis means a call for energy substitution. In consideration of resource prices skyrocketing, energy depletion, security and environmental problems that are caused by fossil fuels, it is urgent to establish a new energy system for the whole world, whether from the perspective of security or sustainable development. As described by Jeremy Rifkin in his book *The Third Industrial*

Revolution, in the near future, each building is a small power station. Everybody is an independent producer of green energy. In addition, people can upload the surplus electricity onto the grid. That is what I have often said—"self-sufficient with the surplus pooled into the grid."

Admit it or not, mankind will step into a "post-carbon" age. The integration of information technology and renewable energy will herald a brand-new Industrial Revolution.

> Following the process of the third Industrial Revolution, new energy is not supplemental, but substitutive, for fossil fuels. By 2035, clean energy will account for 50% of the world's total consumption of primary energy. The age of large-scale substitution of new energy for fossil fuels has begun!

In this revolution, renewable energy will be the key. One essential element of each Industrial Revolution—energy substitution—is very likely to be restaged. New energy will replace traditional energy.

Coal and petroleum represent traditional energy. What is going to represent new energy?

The Photovoltaic Revolution: The "Title Song" of the New Energy Revolution

Why is a French farmer willing to invest 20 million Euros[1] to build a solar power generating greenhouse? Why didn't Zhao Chunjiang, "the first person practicing rooftop solar power generation in Shanghai," not mind waiting 7 years, before finally incorporating his surplus electricity into the national grid?

"Solar power, unlimited amount, endless coverage, non-stop illumination, impartial coverage, free acquisition, and pollution-free consumption." In the third Industrial Revolution, photovoltaic is the optimal choice of renewable energy. After being developed for three generations, three prerequisites have been put in place for the large-scale development of photovoltaic.

The Good Show Is Unfolding

In 2008, at the Vosges in eastern France, a farm with planting and breeding animals as its main businesses drew lots of attention from other farmers and many investors. Westphal, the farm's owner, built a huge solar power generation device on the rooftop of his five big greenhouses. It is 36,000 m² large with an installed capacity of 4.5 MW.

Westphal's project budget was 20 million Euros. The project can supply electricity to 4,000 households after completion, and is expected to realize annual revenue of 2 million Euros. Because he got the loan from Banque Populaire and Credit Agricole, he also signed a power supply contract for 20 years with the government. Westphal's new means of income generation is the envy of other farmers. According to Taiwan's *United Daily News* report, on February 25, 2009, due to the drop in the price of farm crops and the rise of agricultural costs, the average revenue of French farms in 2008 fell 15%.

In China, similar stories unfold occasionally. On December 26, 2012, Xu Pengfei in Qingdao made a reality of his idea of a self-built distributed PV system for home use, which has an installed capacity of 2kW, with a grid voltage of 380/220V. The spare electricity after home use is incorporated into the grid. This PV system cost an investment of only about 20,000 yuan, generating 8.5 kWh on the day of being incorporated. It is expected that the annual power generation will exceed 3,000 kWh.

Compared with Xu Pengfei, Zhao Chunjiang, Director of the Solar Energy Research Institute in Shanghai University of Electric Power and "the first person practicing rooftop solar power generation in Shanghai," gained a hard won victory. Since the end of 2006, Zhao Chunjiang built up a family-type rooftop solar power station out of his own pocket, which has run continuously for 7 years, generating nearly 9 kWh per day. After supplying electricity for his family, the one-third leftover wasn't incorporated into the State Grid. On the last day of July 2013, Zhao Chunjiang finally was visited by staff from Shanghai South City Electric Power Company, and signed a power purchase agreement with the company: the excess electricity would be uploaded onto the grid. The electric company temporarily purchases the surplus at the price of 0.477 yuan/kWh.

Zhang Changqi, a farmer from Taixing City in Jiangsu Province, spent 20,000 yuan for a small PV power station on his rooftop, which can supply the power for 2 to 4 households. He has submitted applications to the Taizhou Power Supply Company for being incorporated into the State Grid.

Gao Song, a citizen of Wuhan City, built a small power station consisting of 18 solar panels. Currently, it's been grid-tied, and expects to save him nearly 2,000 yuan per year.

This kind of story will become more common. And the small home solar power plant is no longer an isolated island of energy in China.

This kind of story will further substantiate the coming business model of renewable energy. As Jeremy Rifkin said, the second economic pillar of the third Industrial Revolution is to transform buildings in every continent into micro power plants, to collect renewable energy on the spot.

Making Its Way from the Oil Crisis

After the substitution of coal for firewood in the first Industrial Revolution, and the substitution of petroleum for coal in the second Industrial Revolution, a new energy is bound to replace petroleum in the third Industrial Revolution.

So-called "new energy" cannot be understood as a "new" "energy revolution." A more accurate description would be a revolution of "new energy." We must focus on the concept of "new energy," so as not to deviate from it. Only by seizing its essence can we have a good understanding of the third Industrial Revolution. Otherwise, no matter how much is written, the point may be lost.

So, what is the leading candidate to substitute for petroleum?

Solar energy is the most advantageous option among various renewable energies. To explain its benefits, I made up a doggerel: "Solar power, unlimited amount, endless coverage, non-stop illumination, impartial coverage, free acquisition, and pollution-free consumption."

First, seen from the angle of energy, the sun illuminates every corner of the globe. People of different races, nationalities, and religions all enjoy the sunshine. The amount of solar energy arriving on Earth per second is equivalent to the heat generated by burning 5 million tons of coal. This means that it takes only one hour for the Earth to receive an amount of solar energy which equals the amount of energy that the Earth needs per year. Second, from the angle of "non-pollution," solar power has a prominent advantage among the new energies. What's more important is that solar energy has neither potential risks nor uneven distribution. Furthermore, it's not constrained by other resources. So, no matter if seen from the perspective of natural endowment or real use, only solar energy can be the representative of the new energy age and the main force to substitute for fossil fuels.

Like any other new energy, exploitation of solar energy has experienced a long course, stimulated by the ups and downs of petroleum. The development of solar power stagnates when oil prices go down, and accelerates when prices go up.

People have been using solar energy for more than 3,000 years. But the history of using it as a kind of energy and dynamic power is only 300-odd years old. In 1615, French engineer Salomon de Caus invented the world's first engine driven by solar power. Its operating principle was to collect sunshine by means of condensed light and use solar energy to heat air to make it expand and pump water.

The breakthroughs in utilizing solar energy are mainly achieved in the 20th century. From 1900 to 1945, research in solar energy showed slow progress, because fossil fuels were developed in great quantity, and petroleum was cheap and rich. During the 20 years after World War II ended, people of insight noticed the soaring consumption of oil and gas, quickly depleting the reserves. They called for pushing forward research in solar energy. At that time, the basis of large-scale application of PV power generation—the silicon solar cell—was successfully developed by the Bell Laboratory in America. Ten years later, Israel Talwat and other people raised the basic theory of selective coating at an international solar heat conference, and developed practical black nickel selective coating, creating conditions for the development of high-efficiency heat collectors.

Solar power utilization didn't get much attention until the oil crisis in 1973. OPEC applied methods such as production reduction, price raising, etc. to support people in the Middle East and to protect their own interests. This made the economies which imported cheap oil suffer severely. Governments realized that oil had become a key factor deciding a country's fate. Oil is also an extremely unstable element. The current energy structure must be changed.

As a result, people began to see solar energy as "much-needed supplementary energy for the near term" and a "basis of the future energy structure." The research and development of solar energy entered the fast lane.

In 1973, the U.S. made a federal-level sunshine power generation plan, increased solar power research grants sharply, and established a solar power development bank to promote the commercialization of solar power products. In 1974, the Japanese government released the "Sunshine Project" in which related R&D projects included solar house power, industrial solar energy systems, power generation by solar heat, solar cell production systems, large and dispersive-type photovoltaic power generation systems, etc. In China, "The First National Conference for Exchanging Work Experience on Solar Power Utilization" was held in Anyang, Henan Province in 1975, pushing forward the development of solar energy. During the same period, the CPC (compound paraboloid collector), evacuated solar collector tubes, the amorphous silicon solar cell, photolysis of water for hydrogen technology, and the solar thermal power plant, were successfully developed.

However, from 1980 to 1992, the solar power tide fell again, due to the slump in oil price while the price of products in solar power remained high, with no major technological breakthrough. In 1992, the UN World Environment and

Development Conference was held in Brazil and passed a series of important documents such as the Rio Declaration, Agenda 21, and the UN Framework Convention on Climate Change. It established the sustainable development model, and incorporated the environment and development into a unified framework. Solar power didn't receive great attention until then.

In 1996, the UN held the World Solar Energy Summit in Zimbabwe. Important documents were discussed, such as the World Plan of Action (1996–2005), the International Convention on Solar Energy, the Strategic Plan for World Solar Energy, etc. The Harare Declaration on Solar Energy and Sustainable Development was released after the conference. This conference further demonstrated the determination of the UN and various countries to develop solar energy, requiring global common action to widely use solar energy.

Three Prerequisites Are in Place

The solar power era has finally come into the 21st century. Although the sun doesn't shine on us all the time, and we are not currently able to collect and make use of all of the solar energy, humanity can realize solar power application at a reasonable cost with the technologies that we already have.

At present, there are two methods of use for solar power. One is to generate electricity with the heat from solar radiation, i.e., the "photo-thermal mode," whose rationale is to use accumulated sunshine to heat a medium (liquid or gas) in a heat collector, then transform that thermal energy into mechanical energy, and then convert mechanical energy into electrical energy.

The other method is the "photovoltaic mode" whose full name is the "solar energy light volt generating electricity system" (also called the PV system). It is a kind of photovoltaic effect that utilizes semiconductor materials in solar cells to convert solar radiation energy into electric energy directly. A single solar cell can only generate a small amount of electricity. But a series of parallel cells can constitute a solar cell array with high output power.

Photo-thermal power generation and photovoltaic power generation have their own strong points and weaknesses. The main problem with the former is that its cost can drop sharply only through the effects of scale, which means that it must be used mainly in large-scale power station construction. The advantage it has of storing energy and its rationale similar to thermal power make the grid more receptive of its electricity. But it has the disadvantage of energy loss caused by long-distance transmission, which makes this mode not quite competitive compared with conventional energy.

The size limitation in photovoltaic power generation is relatively small, making this mode more applicable to small power stations and distributed power stations. For instance, in places where no vast amount of vacant land can be found to build large-scale solar power stations, we can build photovoltaic power generation systems that are grid-connected or off-grid attached to buildings.

With the background of a growing Chinese and global solar power generation market, these two kinds of technical route have their own markets. But seen through Rifkin's description of the third Industrial Revolution, photovoltaic power generation is undoubtedly more suitable for future independent and distributed production.

All the photoelectric converting in photovoltaic power generation has already been completely included in a module with functional independence. Therefore, it is very suitable for distributed power generation. Concentrated solar power generation in photovoltaic power generation is also based on the overlapping effect of a number of solar cell modules. It is assembling and connecting single cells.

To be sure, to realize large-scale photovoltaic, three basic factors are needed: conversion efficiency, grid parity and storage. Currently, these factors have been put in place.

The core technological indicator of conversion efficiency is "photoelectric conversion efficiency," i.e., the ratio of electric energy to the luminous energy per unit which is converted to it. Photoelectric conversion efficiency is not only a technical indicator, but also an economic indicator determining the utilization of PV. The higher the efficiency, the lower the production cost of the electric energy per unit, and the wider the utilization of PV power generation.

After the first solar power station was built in France in 1969, progress in PV has not been as fast as people thought it would be, due to the constraints of low conversion efficiency and high cost for quite some time. Someone immediately brings up "high cost" as soon as PV is mentioned.

But things are always changing. People often overestimate change over one or two years, and underestimate change over 5 to 10 years. As a result, it's impossible to see the trend in development. Changes in one or two years are still *quantitative*, while the accumulation of quantitative changes in 5 to 10 years may lead to *qualitative* change. Moore's law is on stage in the PV industry.

Ten years ago, the photoelectric conversion rate was less than 10%. At that time, people thought it quite remarkable to reach 10%, or 11%–12%.

This stride forward took a long time. But reality quickly goes beyond expectation. Currently, thin film's photoelectric converting rate has reached more than 17%. And it is accelerating. In the past, it took at least 2 years for the conversion rate to increase by 1%. But things are totally different now. During the less than 5 months from May 2012 to September 2012, MiaSole Company raised the conversion rate from 13.4% to 15.5%. At present, the highest R&D conversion rate has reached 17.6%.

It's more mature for large-scale PV application to be integrated into the power grid. The production of inverters has become a key component of the PV industry. Inverters can convert the direct current released by PV to alternating current. And breakthroughs have been achieved for technical difficulties. It's been technically feasible for quite a long time to transform the current grid of "unified power generation and supply" into distributed power stations which feature "self-sufficient power generation with the surplus going to the grid." And the business model and policy are taken into consideration now more by PV practitioners.

Meanwhile, the smart grid age may be just around the corner. The so-called smart grid is the rationalization of the power grid, and is also named "Grid 2.0." It is based on an integrated and high-speed two-way communication network. It realizes the goal of making the grid reliable, safe, economical, highly-effective, environmentally friendly and safe in use by the application of advanced sensing and measuring techniques, equipment technology, control methods and decision-support system technology.

In 2005, the European Technology Platform (ETP) for the Electricity Networks of the Future was set up, whose purpose is to study how to transform the grid into a service network of interaction between users and operators, so as to improve the efficiency, safety and reliability of power transmission and distribution systems, and clear barriers for the large-scale integration of distributed and renewable energies in Europe. In 2009, the smart electric meter was first applied in countries like Germany, France, Italy and Spain. The EU also has a super smart grid plan, whose goal is to combine high voltage transmission with a smart grid.

Currently, the storage situation is manageable. Due to the volatility of solar power generation (in the daytime, sufficient sunshine brings big generating capacity; at night power generation is impossible). We need the storage battery to store electricity to adjust for this fluctuation. Storage batteries with a small volume have been widely used and improved continuously. Urban lighting by solar power and the battery used in the electric automobile have become reality. There is no barrier to the electricity storage equipment needed by the family-type solar power station. Though currently there is no real breakthrough in large storage technology that can be used repeatedly, this doesn't affect getting PV power into the grid fundamentally.

Currently, the grid's biggest peak-shaving occurs between day and night. Most industrial corporations consume electricity in the daytime, which coincides with the power generation peak period of solar power stations. Thus storage pressure is reduced as well as the proportion of the external power supply, which alleviates the peak-shaving pressure on the grid.

In the long term, electricity storage not only better resolves the peak-shaving problem in the system, but also better resolves the problem of mobile utilization of electricity. This is a challenge not just for the PV industry, but for the whole electric power industry and the fields of science and technology, and it needs the attention of all of society.

The third point raised by Jeremy Rifkin in his five economic pillars of the third Industrial Revolution, i.e., use of hydrogen and other storage technology to store intermittent energy in all buildings and the infrastructure, is quite feasible in the future.

Third-Generation Technology Has Arrived

Beside the aforementioned advantages, solar energy has another technical advantage, i.e., solar power generation technology is continuing to mature. Solar power's cost has been rapidly approaching the cost of traditional power generation.

> The solar photovoltaic industry is the most promising clean energy industry. Thin film and flexibility are a general trend of global solar power development. I believe that 0.5 yuan/kWh cost of PV power generation is a goal for the development of new energy. Because the cost of thermal power generation (including the environmental cost) has exceeded 0.5 yuan/ kWh and continues to grow, the age of large-scale substitution of new energy, particularly solar power, for traditional energy has come.

Solar PV generation technology is mainly reflected in the solar cell, which is also called the "solar chip" or "PV cell." It is a kind of PV semiconductor wafer which uses sunshine to generate power directly. As long as it is exposed to the sun, electric current can be put out immediately.

In 1883, the PV cell was first successfully made by the American inventor Charles Fritts. His method involved coating the surface of selenium with a thin layer of gold. The highest photoelectric efficiency of this kind of PV cell is less than 1%. However, during the next several decades, solar cell technology has evolved in three generations, which cannot be considered as in the same league: from the silicon substrate of the first generation to the thin film of the second generation, and then to the composite thin film material of the third generation,

The hull cell is worth mentioning. Its biggest advantage is cost. With the same area, the thickness of crystalline silicon solar cells is about 0.2 mm, while the thickness of the hull cell is only 1% of that, minimizing material consumption. In terms of the use of silicon material, the unit cost of the hull cell is far lower than that of the crystalline silicon solar cell.

So far, the solar cell slice has reached the 0.1 mm level; the highest R&D conversion rate of the hull cell has reached 18.7%. There have been cell slices with a conversion rate of 19%–20% in the mono-crystalline silicon market, and around 18% in the polycrystalline silicon market.

Currently, due to new technology and economy of scale, the cost of PV power generation is dropping 8% per year, drawing more and more attention from many countries. This is related to the policy of the "feed-in tariff" implemented around the world.

With the wide use of photovoltaic technology and the support of government policy, tens of thousands of small PV energy producers have been incorporated into the smart grid. This demonstrates that the integration of new energy and new technology, predicted by Rifkin, has been realized. The new social reform phase has begun.

Rise of a Big Power

Two hundred years ago, with the first Industrial Revolution, the UK emerged as a great power through the replacement of wood with coal for energy. However, China, indulging in the prosperity of the Qing Dynasty (from the reign of Emperor Kangxi to Emperor Qianlong), turned a blind eye to all this, thus losing the opportunity to lead the world. One hundred years ago, when the second Industrial Revolution broke out, oil (the substitute for coal) enabled the US to become a powerful country, while China was attacked by foreign powers and trapped in war.

Now, the 21st century has brought us the third Industrial Revolution, in which renewable energy is going to replace fossil energy. This time, can China grasp the opportunity and light up the "Chinese Century" with solar power?

From Kaifeng to New York

On May 22, 2005, an article with the title *China, the World's Capital* or *From Kaifeng to New York, glory is as ephemeral as smoke and clouds**) unexpectedly appeared in the Opinion section of the *New York Times*. The author, Nicholas D. Kristof, is an old China hand who used to be a permanent reporter for the *Times* in China.

In May 2012, visiting China again, Kristof was amazed by China's fast development. He published a series of articles in his column talking about China, including the commentary focusing on Kaifeng and New York, an old and a new world capital.

The piece sought to remind Americans: China is reviving, and in the next few decades it will surpass the US as the largest economy in the world, so the American people should stay alert.

Entering the 21st century, the US remains the only superpower in the world, with New York being one of the most important cities around the globe. However, 1,000 years ago, the "capital" of the world was Dongjing, the capital of China in the Northern Song Dynasty, today known as the city of Kaifeng. The once-prosperous Kaifeng was captured perfectly in "Riverside Scene at Qingming Festival," a painted scroll by Zhang Zeduan, a famous painter from the Northern Song Dynasty (960–1127 a.c.e.). In its heyday, Kaifeng was home to a population of more than 1,000,000, while the total population of London at that time was only 15,000 – and New York did not even exist yet.

Kristof said, "If you look back at the sweep of history, it's striking how fleeting supremacy of a country is, particularly for individual cities."

One thousand years of historical changes turned the positions of Kaifeng and New York upside down. What made Kaifeng, a world-class metropolis, decline? And what made China, whose GDP was once one-third of the world's, fall behind? Industrial Revolutions are the answer to both questions.

Missing "the First Industrial Revolution"

In the middle of the 18th century, the scarcity of wood, the world's major source of energy, drove the British to decisively turn to harnessing thermal and

---

* Kaifeng was one of the capital cities of ancient China

mechanical energy. By adjusting its energy structure, the UK wholly transformed its industrial system and scored a critical breakthrough in economic history, thus igniting the first Industrial Revolution.

If we could go back to the UK at the beginning of the 18th century, we would see numerous textile mills lining the banks of clean rivers, and these rivers slowly drive the old water wheel with a rhythmic cracking sound.

In 1764, a weaver named James Hargreaves invented the spinning jenny, which enabled yarn making with multiple spindles and thus improved work efficiency tremendously. As a result, the cloth trading floors of Yorkshire bustled with activity, selling cloth pieces as thin as cicadas' wings. The quality of textiles advanced quickly, leaping ahead due to the invention of spinning mules and automatic spinning machines in the years that followed.

In 1782, the double-acting steam engine was invented. From the second half of the 18th century, steam engines were widely used in production. This was more due to the success of coal power than to Watt's contributions. Because coal is easy to carry and has higher thermal efficiency, any plant or mine could use the new machine without the necessity of building the workshop near woods or a river.

Machine-based industry would not have developed without coal and steel. Coal was the primary energy source of contemporary industry, and steel is the material for making machinery. Because it was rich in these raw materials, the UK traveled very fast down the path to dominating the world.

Steam engines also gave the Britons the confidence to extend shipping routes. Launching big ships fueled by coal, they built colonies around the world and eventually expanded the UK into an empire on which the sun never set.

The One Who Owns the Resources Is One Step Ahead. Although English clerics had burned coal to warm themselves as early as the 9th century, most Britons were indifferent to coal until the 17th century. This is because the UK was endowed with large forests, enough for smelting iron and heating. For example, the air furnace that once prevailed in the UK in the 15th century was fueled with charcoal. An iron mill consumed more than 1,036 km^2 of forest every year.

According to John Neff, a financial investor, 2,000 carriages of lumber could only meet the demands of the wine-making industry in London for a year in Elizabeth's reign, and till the reign of King James I, a glass-making workshop needed 4,000 carriages of lumber each year. Besides, "the navy also needed wood to build warships." Moreover, only premium timber was accepted and so wood was consumed in large quantities.

Mass lumbering made industrial production and the lives of the Britons subject to the energy supply. From 1500 to 1630, the price of firewood rose 7-fold, while that of commodities during the same period only tripled. In the 17th century, soaring urban populations led to a surge in demand for wood for construction, heating and handicrafts. Plus, with the advent of the so-called Little Ice Age, winters in Britain became much colder and longer. The British had to seek alternatives to wood, and that was why coal began to be mined on a massive scale. Statistics show that by 1700, coal production in the UK had reached 2.5 to 3 million tons, 5 times the aggregate coal production of the rest of the world.

Since coal mines in Britain, the Pas-de-Calais region in France, and the Ruhr region in Germany were constantly developed, France's coal production grew from 4.4 to 13.3 million tons from 1850 to 1869, and German coal production rose from 4.2 to 23.7 million tons in the same period. The world's coal use against total energy consumption jumped from less than 30% in 1830 to 48% in 1888. In 1920, coal consumption made up as much as 62% of the total.

From the 13th century onwards, when the Song Dynasty of China had just begun, the world entered a "coal era" one country after another.

What did China do when all of Europe experienced such complete transformation?

China was still evolving from an agricultural into a modern civilization then, relying mainly on land. A big increase in grain production can be regarded as the biggest change, if any, for China at that time. And that was also because maize and sweet potatoes had been introduced into China thanks to their development in America.

But China was then still the largest economy in the world. For example, in 1800, China's economic aggregate accounted for 30% of the world total, around 6 times that of the UK. China also made up 33% of the gross global manufacturing output, and had a soaring population.

When the first Industrial Revolution took place, the Qing Empire was at its peak. In the 42nd year of Emperor Qianlong's reign (1777 a.c.e.), the imperial court reiterated that firearms were not allowed to replace swords and knives in the imperial examination of martial arts. However, 5 years later, James Watt successfully upgraded his steam engine.

In 1793, Emperor Qianlong met George Macartney, British ambassador to China, in his Mountain Resort, claiming, "My great empire possesses everything. We don't need to import goods and we don't need foreign trade." Nevertheless, the British mission wanted to open the Chinese market, to explore the Far East market for British capital and maximize its potential.

Macartney's mission to China ended up in failure with the trade imbalance between China and Britain remaining unresolved. Therefore, British merchants blazed a trail with opium, and the First Opium War took place.

In the later part of the prosperous days from Emperor Kangxi (1662–1723) to Emperor Qianlong (1736–1796), China not only lacked the kind of global vision that was necessary, but also paid no attention to what was happening in the West. Because of its arrogance, its seclusion policy, and its suffering from a series of blows both at home and abroad, China missed the opportunity to participate in the first Industrial Revolution.

Due to the first Industrial Revolution, countries with smaller populations and fewer workers for the first time increased their share of global manufacturing output.

So let's have a look at the change of proportions held by China and the UK in manufacturing. In the centuries before 1840, China had always been the largest country in terms of manufacturing industry. Around 1840, the productivity of British factories increased significantly, boosted by the first Industrial Revolution, and helped Britain replace China as the world's largest manufacturer.

Putting more attention on ethics than on science and technology, on agriculture than industry and commerce, on self-seclusion than communication with the world, China, still relying on firewood, finally lost the competition with the newly coal-fired Britain, because China missed this first Industrial Revolution.

Letting the "Second Industrial Revolution" Slip By

The transformation from the coal era to the oil era marked fossil fuel usage evolving from its first phase to a second one. Thanks to the application of electric power, the invention of the internal combustion engine, the development of new vehicles and the advancement of new means of communication, the four pillar industries (namely energy, iron and steel working, petrochemical technology, and automobile manufacturing) came into being. As a result, industry began to take a dominant place in national economy.

In a manner of speaking it was the first Industrial Revolution 200 years ago that established the rudiments of the world market, while it was the second Industrial Revolution 100 years ago that formed the ultimate worldwide market.

The energy revolution, along with the upgrading of smelting, electric power and communication technologies brought about by the industrial revolutions, together with powerful machines, enabled people to get access to each and every corner of the globe. Colonial expansion and war became major motivations of those emerging powers, contributing to the origins of World Wars I and II.

The first Industrial Revolution was one of the triggers for the American Civil War, which in turn paved the way for the second Industrial Revolution in the US. In addition, America's significant combat gains from World War II and its intact industrial, scientific and educational systems enabled the US to become a new empire, ousting the UK from its top spot.

In 1894, the US ranked top of the world in terms of GDP, as the most powerful economy in the world. It was only 118 years since this new country had been founded and only 400 years since this new continent had been discovered.

But what was happening to China in the late 18th century?

At that period of time, China had not overthrown the rule of feudalism and that old and outdated political system still constrained productivity. China did not enter the electricity and oil era, nor host its own first Industrial Revolution. Suffering from the Sino-Japanese War of 1894–1895, the Hundred Days' Reform, the Boxer Uprising, the Siege of the International Legations, the 1911 Revolution, and warlord in-fighting ... China was trapped in turmoil and turbulence.

China did go through a so-called golden decade, but the industries introduced into China, e.g., the booming silk-reeling industry in the south in the middle and lower parts of the Yangtze River, were just the echoes of the first Industrial Revolution in developed countries. Unfortunately, no complete industrial system that might lay a foundation for the second Industrial Revolution was established in China, so its status as an agriculture-dominated country was not fundamentally changed. Even worse, its struggles with industrialization were further interrupted by the Japanese invasion.

The closed-door policy and the multiple wars of the time distracted China from participating in the second Industrial Revolution. And a direct consequence of that was that the economic aggregate of China, once a world leader, dropped to one-twentieth of the world's total by 1950.

From 1820 to 1913, the economic aggregate of the world quadrupled, with that of Western Europe growing to 6 times its size, and Britain's increment to even more than 6 times. However, intruded into and exploited by imperialist countries, the total Chinese economy registered negative growth in the first 50 years of this period, and small growth in the second, ending up with almost no growth at all.

This is how China let the second Industrial Revolution slip through its fingers. As opposed to merely missing the first one, the Chinese faced more than just being outrun by others this time. The Chinese people felt the pain of total failure, and learned in depth the lesson of "being vulnerable to attack if the country lags behind."

> Scholars often maintain that technological revolution is the core of Industrial Revolutions. However, if we go a step further, we will see that the first two Industrial Revolutions in history were driven by and centered on "energy substitution." And the boom and bust of powerful countries hinges on whether they use new energies.

The last century can be seen as one dominated by the US, but it was also the century of oil.

As early as the late 19th century, the US had caught up with and outpaced the UK. In 1871, steel production in the UK was 6.6 million tons and that of the US was 1.7 million tons; in 1900, Britain's steel production was 9.1 million tons while America's had soared to 14 million tons. In 1850, the industrial output of the former accounted for 39% of the world's total and America only made up 15% of that pie; but then by 1913, British industrial output took up only 14% while the US got a far bigger share—to the tune of 36%. In other words, from 1870 to 1913, the industrial production of the US grew 81 times while that of the UK grew by only 13 times. In 1913, the US became the largest contributor of industrial output in the world, and the UK fell to third place.

Based on this data, the American economy reached global supremacy in the 1980s. Jeremy Rifkin believes that the "heroes" behind the American success were abundant and cheap oil, cars and people's avid consumption. For the last two oil was a prerequisite.

Today, China is standing at the same starting line with its peers for the third Industrial Revolution. And the new energy industry is developing so fast that no one could have imagined it only a few years ago. For China, this is a once-in-a-blue-moon opportunity. If grasped, it can help China redefine its position as a great power in the world.

How Much Longer Can the "Oil Century" Last?

It's fair to say that during the second Industrial Revolution, the fundamental development pattern for countries was a synchronized and fast increase in both the country's economy and its consumption of fossil fuels.

From 1950 to 1973, the global economy grew at an unprecedented, record-breaking speed of 4.91% on average, almost 2.5 times the growth rate in the golden age of coal (1870–1913). The global economic aggregate quadrupled. The Japanese economy grew by an amazing 9.29% in this period, and its economic aggregate increased nearly 8-fold; the American economy increased by 4.8%,

2.2 times that in the previous century, and its economic aggregate grew three-fold. The energy powerhouse behind such growth is abundant and cheap oil. By 1973, oil production accounted for 43% of total energy production, giving birth to an economic system and civilization entirely dependent on oil.

In his book *The Third Industrial Revolution*, Jeremy Rifkin proposes a shocking argument from the outset: the international financial crisis that broke out in July 2008 could be taken as the peak of globalization. Governments tried their best but still could do nothing about it, because we are in an economic system that depends on oil and other fossil fuels. Accepting that premise or not, we are now in the last phase of the second Industrial Revolution and the Oil Century.

As an executive working in the new energy sector, I felt quite delighted, as if I had met a soulmate, rather than shocked, when I read this argument. In my opinion, oil is going to leave the arena of history much sooner than people think, and if China can catch the third Industrial Revolution, it's very likely that the oil system dominated by the US will be replaced by a solar power system led by China.

So then, how many more years can the glory of the Oil Century last?

First of all, let's look at proved oil reserves. Based on the data published by the International Energy Agency (IEA), the volume of proved oil reserves (PORs) kept growing from 1991 to 2011. Notably, from 2001 to 2011, it grew by more than 5 billion tons per year. At present, the world's PORs add up to approximately 1.652 trillion barrels.

Next, let's move on to oil production. King Hubbert, a geophysicist with Royal Dutch Shell, once made a prediction in the 1950s that global oil production would reach its peak (70 mbd) around 2025–2035. And according to IEA's *World Energy Outlook 2010*, that peak had already been reached in 2006, and future energy exploitation will be increasingly difficult and costly, leading to global economic fluctuations. As a matter of fact, the world oil production in 2008 registered 81.73 mbd (7 barrels=1 ton).

Next, let's talk about oil consumption. Nowadays, the world consumes 80 million barrels of oil every day. From 1999 to 2006, global oil consumption increased by an average of 1.68% per year. Around 40% of the oil was burned by motor vehicles, whose number in use globally was 662 million in 1999 and 847 million in 2009, jumping by 2.49% annually. The oil self-sufficiency rate of most countries dropped rapidly while the resource was almost entirely controlled by the six Gulf countries with a population of less than 0.5% of the world's total.

The advancement of exploration technology, the increasing investment in finding new oil sources, and a different statistical methodology have boosted

POR increase over exploitation activities. Nevertheless, oil is a non-renewable resource whose total reserve on the earth is a definite figure, and someday the reserve will be exhausted. At the current exploitation rate, oil reserves will only last for another 41 years.[2]

Rapidly-declining oil resources, unbalanced oil distribution and the monopoly of interest groups are very likely to trigger a disaster in the last phase of the Oil Century, e.g., some countries may wage war for oil.

We can get a rough idea about this from the previous oil crises.

In October 1973, when the Yom Kippur War broke out, the oil-producing countries in the Middle East punished the US and other countries for supporting Israel by cutting production, banning oil export to some of these countries, and raising oil prices for others. Such moves led to the first oil crisis, and gave birth to the IEA – a US-led organization which initially contended with OPEC, and then shifted to dialogue and cooperation with the latter.

The Islamic Revolution of Iran (from 1978 to February 1979) and the subsequent Iran-Iraq War (from September 22, 1980 to August 20, 1988) resulted in soaring international oil prices, thus triggering the second oil crisis. Consequently, the Carter administration promulgated the National Energy Act (NEA) and the Crude Oil Windfall Profit Tax Act among others, which indicated that energy conservation and development of alternative energies were already on the agenda. Moreover, every country started to attach importance to strategic oil reserves.

Rising oil prices caused by the Gulf Crisis (1980s) and the Gulf War (1990) gave rise to the third oil crisis. Its direct result was the US Congress passing of the Energy Policy Act. After the breakup of the Soviet Union, the abundant oil source in the Caspian Sea area started to demonstrate its significance in the international arena, becoming a key component of the American strategy of energy diversification.

Ushering in the "Chinese Century"

After his bestseller *A Century of War: Anglo-American Oil Politics and the New World Order*, Frederick William Engdahl, a renowned American economist and geopolitical analyst, produced another book, *The War in Mali and AFRICOM's Agenda: Target China*. He revealed that the US had been lying about oil depletion in order to control oil and exert hegemony in the world with oil under its control.

The US remains one of the countries with most PORs in the world by far. Perhaps just as William Engdahl pointed out, in the future, America will continue to exert world hegemony with oil. On the other hand, decision-making in energy development in the US government will be influenced by the clout of large energy companies. It's therefore possible that the US will lose the chance to lead the world in a new round of competition, because its oil-based political landscape may act as a shackle for it.

In the new energy sector, this likelihood is gradually turning into reality. At the beginning of his first term in office, President Obama identified new energy as a key industry to develop and provided loan guarantees to some energy companies. Solyndra, a solar panel manufacturer based in California and one of the beneficiaries, received a loan of USD 535 million in 2009. However, because of poor management, the company declared bankruptcy in late August 2011 and laid off all its 1,100 employees.

This case was used by pro-oil Republicans as evidence to refute Obama's claims. For example, in the 2012 presidential election, the Republicans accused the Obama administration of providing loan guarantees to Solyndra out of a political motivation and bringing about huge losses for taxpayers. Facing such attacks, the Obama administration is more cautious in its support of clean energy.

> China lagged behind in the previous two Industrial Revolutions. So, in this PV (photovoltaic) revolution "China should take the lead" and contribute to the development of clean energy in the world. If China succeeds, we will have contributed our efforts to China's revival, as well as to peace and development in the world.

In contrast to the US, China has made firm decisions and consistent policies to promote clean energy. For example, the Chinese government launched a "Rooftop Solar Power Program" and the "Golden Sun Solar Power Demonstration Project" in 2009. And in 2012, when the PV industry was on the brink of an abyss, the Chinese government decisively promulgated a series of policies to support and subsidize the PV industry.

In the first half of 2013, the Ministry of Finance issued a *Notice on Subsidies for Distributed PV Power Generation Based on Quantity of Power Generated & Other Issues*. The policy that PV power station operators are entitled to 50% VAT (Value Added Tax) refunds when tax is collected was released in September of the same year. The VAT rate for these operators has been lowered from 17% to 8.5%, which is equivalent to raising on-grid power tariffs and the ROI (Return on Investment) of the stations. That is definitely good news for the operators.

I believe that the Chinese government made the right decision. As I have suggested, the future may hold the co-existence of multiple new energies, but it is only solar power that can save the day.

Energy substitution will come at a critical moment, and so will sustainable development in China. The question is: what role will China play in the new energy revolution, as a follower or in the vanguard?

Statistics have shown that China will catch up and surpass its competition. For instance, China has maintained an annual GDP growth of more than 9% for 30 years, which makes it the fastest-growing economy in history. In 2011, China's GDP was USD 7.3 trillion, its gross volume of exports and imports accounted for around 10% of the world's total. Moreover, China is the third greatest scientific and technological power in the world, as the home to around 926,000 researchers. China has already joined the ranks of the most advanced elite in many sectors.

The National Situation Analysis and Research Group of the Chinese Academy of Sciences (CAS) forecasts that from 2020 to 2030, China's economic aggregate would be the largest in the world; in the period 2040 to 2050, China's per capita GDP would reach the current level of developed countries; and by the end of 21st century, both China's per capita GDP and per capita level of social development would be at the same level as that of developed countries.

This means that the "Chinese Century" is around the corner. But can such a period be sustained by China with such fast growth, given its current energy, capacity, resources and the environment? I think that if it can, there is only one way to get there. China has to rise in a green way, gaining maximum economic and social benefit, minimum environmental cost and the most reasonable consumption of resources.

In this way, the "Chinese Century" has to be a "century of solar power."

Who Will Lead the Ultimate Replacement?

On this planet we live on, humanity has given birth to its 7 billionth baby, and her name is Danica Camacho. How can we ensure that Camacho and succeeding generations can live a green, peaceful and sustainable life with new energy? The answer is solar power. Because fossil fuels have an Achilles' heel: "quality, quantity, and distribution," which render them unable to support sustainable development in the world. Unfortunately, wind, nuclear and tidal energies also have their limitations. So the only reliable energy substitute for coal and oil is solar power.

This will be the "ultimate substitution," the reality of which has become more and more obvious to me.

The Achilles' Heel of Fossil Energies

On October 31, 2011, Danica Camacho was born in Manila, in the Philippines. Once she came into this world, her name immediately went down in history, because she was the 7 billionth human inhabitant of planet Earth.

Danica had no idea that her birth was a new milestone in human population. However, I, living far away from her in Beijing, felt concerned. The world's growing population, which has reached 7 billion, will definitely consume more energy. To meet such a demand, what can I do as a corporate executive committed to promoting green energy?

It occurred to me that we should help people understand the inevitability and necessity of the substitution of renewable energies for traditional fossil fuel and urge them to make it happen as soon as possible with their concerted efforts.

Hence, I'd like to make an appeal to the world: the era of solar power has come, and fossil energy is going to rapidly leave the historical arena. This is both an irresistible trend and an irreversible journey. It is solar power that will illuminate the future of mankind!

My belief in the inevitable revolution of new energy has two bases: first, the traditional energy system centered on fossil fuels cannot be sustained indefinitely; second, the new energy system relying on the advantages of renewable energy will become the suitable substitute, step by step. One system cannot last long and the other is certain to take its place. That is a revolution that is sure to unfold.

Why is the traditional energy system based on fossil fuel unsustainable? Because it has bottlenecks and unavoidable disadvantages in quality, quantity and distribution—its Achilles heel.

First, the restraint of "quantity:" the reserve of fossil fuels around the globe is finite. The increase in proved reserves and the exploitation thereof are only a change at the consumption level, rather than an expansion of the whole. As a resource of finite quantity, fossil energy will someday be depleted.

The reserve is limited, while population keeps on growing, along with the world's economy. With industrialization and the improvement of people's living standards, human beings' consumption of fossil fuel is rapidly rising. More importantly, energy consumption and population do not grow at the same pace. The former outpaces the latter several times over.

In 2007, the US Energy Information Administration (EIA) forecasted that global energy demand would be 10.599 billion tons of oil equivalents (TOE) in 2010, 12.889 billion TOE in 2020, and 13.65 billion TOE in 2025. In fact, the energy we consume has already gone far beyond the predicted level.

Second, the quality restraint: Fossil fuels, mainly composed of hydrocarbons or their derivatives, are used through chemical reactions, from which severe environmental pollution and ecological damage may result. Research shows that there are five major pollutants in the air, namely oxynitrides (e.g., NO and NO_2), sulfur oxides (e.g., SO_2), oxocarbons (mainly CO), hydrocarbons (e.g., CH_4) and suspended particles, all of which are emitted in the use of fossil energy.

Using fossil energy directly results in acid rain and ozone depletion. In the past two decades, thanks to fast economic development, environmental damage (especially air pollution) has become a major threat to the survival and general health of people worldwide. The Earth is becoming more and more unfit for people and other creatures to live on.

Thirdly, the restraint in "distribution": fossil fuels are buried deep under the ground. They are only distributed in certain regions and in an unbalanced manner. First, they are unbalanced in quantity. Some countries have plenty of fossil fuels while others have only a little or none. Second, the unbalance in variety. Some countries are rich in oil, some in coals and some others in natural gas. .

Besides the above-mentioned imbalance, supply and demand for fossil fuels is not in equilibrium either. Within a given country, the energy reserves sometimes don't match consumption. On the other hand, some countries hold huge energy reserves while consuming little; other countries consume enormous amounts but only have small reserves. For instance, the Middle East is abundant in unexploited oil, making up almost 65% of the global reserve, but it consumes only a little, given its small population. However, such industry-advanced regions as Europe and America, with relatively low oil reserves, use huge amounts of the fossil fuel, and thus need to import oil from abroad.

Energy is the most critical resource for any country's existence and development, and therefore assuring a stable energy supply is the top priority. In the face of unbalanced resource distribution, countries are inevitably involved in conflicts to match supply with consumption. This may lead to chain reactions that exert an impact on the strategic decisions of countries, and also on the fate of mankind and our civilization.

Which Option To Choose?

So if we want to replace fossil energy, what options do we have? I think solar power is China's optimal choice.

Let's first talk about wind energy. The beauty of wind energy is that it is renewable, inexhaustible and green, and perhaps most importantly it is abundant. Wind is a natural phenomenon caused by solar radiation. When sunlight reaches the surface of the Earth, different parts of the Earth are heated unevenly, thus leading to temperature differences and convection currents in the air. This is how wind comes into being. Although only 2% of the solar power reaching the Earth is converted into wind energy, the aggregate amount is impressive. It is estimated that the wind resource is 10 times the volume of the water resource on Earth, with a power generation capacity of 53 trillion kilowatt hours (kWh) per year. Even if global power demand in 2020 rises to 25–30 trillion kWh, in a purely technical sense we can meet all this demand with only 50% of the wind energy available.

And the cost of wind power generation has fallen year by year to 0.5 yuan / kWh–0.6 yuan/kWh. So wind is currently the most cost-effective renewable resource. For a wind-rich area, wind power generation is more cost-competitive than oil- or gas-fired generation.

China boasts 253 million kilowatts of wind resources based on land areas, and 750 million kilowatts in coastal areas—altogether 1 billion kilowatts. But these numbers only account for the wind energy potential below 10 meters above the ground. If we go further up, to above 50–60 meters, the figure will be at least doubled, to around 2–2.5 billion kilowatts. If two-thirds of these resources get developed, the power generated will be more than 4 times that from water, which means the future development of wind energy is very promising.

However, the distribution of wind resources in China is unsurprisingly uneven. Northeast China, Inner Mongolia, Xinjiang, Tibet and coastal areas are the country's wind-rich regions, while the Southwest and the central areas are relatively short on wind. And wind strength varies significantly in regions with different terrains. In two adjacent regions, wind strength in the region with favorable terrain is usually several or even dozens of times that of the region with unfavorable terrain.

Wind energy development is also hindered by the low density of wind and its unstable strength. What's worse, the power-generating capacity of wind is only one-third that of traditional fossil energy. Long-distance power transmission is pretty costly, which, if included in the total cost, leaves wind energy with no big edge in cost-competitiveness.

Although coastal areas are rich in wind resources, it's much more complicated to develop and operate offshore wind farms than onshore ones. For example, China has not clearly defined the functions of different offshore areas, so the development of wind energy in these regions involves the interests of a variety of different sectors; the ocean and maritime administration, military interests, transportation and fishing departments, etc. If a wind farm sits close to the coast, it is more likely to conflict with the authority of the fishing and environmental protection agencies. But if it is far from the coast, the project may impede shipping routes.

The high maintenance cost of wind turbines also limits the growth of wind energy. Installations far from the coast require the support of special devices and a means of transportation in order to be maintained. Grid integration needs extra investment, and most auxiliary parts are imported. Currently, both the construction cost and operational cost of offshore wind farms exceed those of onshore ones.

Let's move on to hydro-electric energy next. Among the total proved energy reserves in China, raw coal takes up 85.1%, hydropower 11.9%, crude oil 12.7%, and natural gas 0.3%. Among the residual recoverable reserves, raw coal accounts for 51.4%, hydro energy 44.6%, crude oil 2.9%, and natural gas 1.1%. It is clear that coal and hydropower dominate the traditional energy mix in China, and that hydropower plays a significant role, second only to coal.

Although China's water resources are abundant, they are, like wind, distributed in an uneven manner. The volume decreases progressively from the west to the east, with water resources being mostly concentrated in the west and central areas. According to Hanergy's 2006 research, among China's exploitable water resources, East China, Northeast China and North China only owned 6.8%, the five areas in South-central China 15.5%, Northwest China 9.9%, and Southwest China 67.8%. In the southwest region (excluding Tibet), Sichuan, Yunnan and Guizhou provinces held 50.7% of the country's exploitable water resources.

Additionally, there are a higher number of large hydropower plants than there are small ones, and the large ones are concentrated in particular regions. There are 203 large hydropower plants with an installed capacity of over 100,000 kilowatts, and their installed capacity and annual generation capacity account for 80% of the country's total. More than 70% of large-scale hydropower plants are distributed in the four provinces in Southwest China.

What's more unevenly distributed is China's precipitation and runoff, because China's climate is influenced by the monsoon season. Runoff in the 4–5 months in summer and fall makes up 65%–70% of a year's total, while that in winter usually makes up a small proportion. This is why hydropower plants generate a large amount of seasonally dependent power.

And hydropower development also faces challenges. On the one hand, China has a vast population but a limited amount of farmland. Because some farmland has to be flooded in the building of reservoirs, they cannot be constructed in large numbers. By building reservoirs along valleys and rivers, one can reduce such losses, but then it becomes hard to build the high dams that are necessary for such reservoirs. On the other hand, most rivers in China, especially their middle and lower streams, often need to meet the demands of flood control, irrigation, shipping, water supply, fishery and tourism. Therefore, in developing hydropower, planning the projects as integrated wholes is a requirement in order to maximize the economic and social benefits for the nation as a whole.

In the family of renewable energies, biomass energy, tidal energy and geothermal energy have not been utilized in a standard manner and higher quantities and quality are needed. Shale gas, in my opinion, is not a form of renewable energy (see below).

Fortunately, we still have solar power to save the day. The Sun is the major energy supplier for the Earth. Via nuclear reactions, the Sun emits enormous energy. After travelling 150 million km, energy reaching the Earth makes up 1/2.2 billionth of the total, while 30% of that energy is reflected back to the universe by the atmosphere, and 23% gets absorbed by the atmosphere. Eventually, the energy that arrives at the Earth's surface per second can still generate 800,000–850,000 kilowatts of power, equal to the energy released by burning 5 million tons of coals per second. This energy is the source of hydropower, wind power, tidal power and geothermal power. Actually, even fossil fuels are by-products of solar power. That's why solar power is bound to be the best choice in this new round of energy substitution.

The Ultimate Replacement

By making such comparisons, we come to the conclusion that solar power is the best emerging energy source to meet the development requirements of the 21st century. Then when we make solar power generation less costly and grid parity (when the generation cost of solar power is equivalent to the cost of traditional fossil fuels) a reality, distributed solar power stations will get people's attention and rise in popularity. It is at that time that we will witness this new energy's "ultimate substitution" of fossil fuels.

The current situation shows that this "ultimate substitution" is getting closer and closer. In China, there are more and more PV power station projects in the works, and more than 70% of PV capacity is installed in large-scale power stations. Around the world, distributed PV stations make up 68.9% of the total accumulated installed capacity of PV stations. This percentage is over 83% in the US, more than 85% in Germany, and more than 90% in Japan. In 2012, "distributed" PV generation, which is popular in Europe and America, began to prevail in China also. Distributed PV equipment is mainly installed on the rooftops of homes and factories. The power generated is mainly for local use, with excess power being uploaded onto the grid.

Multiple documents issued by the National Energy Administration in 2012 identified replacing large ground stations with "distributed" ones as a policy highlight. The key phrase used in these documents is: "promote large ground-based power stations in an orderly manner... [while] vigorously promoting distributed power stations."

On the morning of October 26, 2012, the State Grid Corporation of China (SGCC) held a news conference, signifying a favorable turn in the policy for grid integration. SGCC declared that it would allow the distributed PV power stations with an installed capacity of less than 6MW to be integrated into its grid. It also promised to facilitate these programs, started to accept and design integration plans, and would charge nothing for the whole debugging process in integration. What's more, the State Council, in its executive meeting of December 19, 2012, announced that China would actively develop its domestic PV application market and endeavor to promote distributed PV power generation.

For China's PV companies, who had then only recently suffered from anti-dumping and anti-subsidy investigations from Europe and the US, the opening of a domestic "distributed" PV market would be an opportunity for them to absorb that PV power capacity.

The policy, showing favor towards distributed PV power stations, has received positive feedback. On July 30, 2013, the first 20MW project of a demonstration program for 400MW distributed PV power generation launched by the Aviation Industry Corporation of China (AVIC) was kicked off at Shijiazhuang Aircraft Industry Corporation Co., Ltd of AVIC. It was the first large project of this category launched after *Opinions on Promoting the Healthy Development of the Photovoltaic Industry* was issued by the State Council on July 15, 2013. By doing so, AVIC also became the first large-scale SOE administrated by the central government to build a distributed PV project on its own rooftops.

The obstacles to lowering costs and raising solar's popularity are expected to be removed soon. Manufacturers of solar cells both at home and abroad are working to cut the cost of such technology, and thanks to their efforts the power generation cost of solar cells is being gradually lowered. According to a seminar on the development of large-scale solar generation co-sponsored by the National Energy Administration and Asia Development Bank, the cost for PV generation is now 50% lower than it was 3 years ago. Without taking land cost into consideration, China's solar price has decreased to below 1 yuan/kWh. At the current pace of development, it's possible that within 3–5 years the generation cost of solar cells will approach the cost of coal-fired power. As a matter of fact, research has shown that the comprehensive cost for such fossil energy sources as coal is more than 0.7 yuan/kWh if environmental impact is included.

In order to popularize PV generation, China has committed itself to working out proper business models to decide on-grid power tariffs, as well as how the government should offer and remove subsidies. After all, it subsidizes PV companies today so that they can survive without subsidies in the future. Such policy issues are in fact being worked on by governments all over the globe.

On July 31, 2013, the Ministry of Finance issued a notice on subsidies for distributed solar PV power generation based on the volume of generated power: the notice stated that the subsidies will be transferred via grid companies to the owners of distributed PV power projects.

Soon 50% of Energy Will Be Clean Energy

Based on my years of experience in the industry, I predict that by 2035, clean energy will make up 50% of the total primary energy usage in the world.

> People may ask, are you a little optimistic? My answer is that if the proportion of clean energy is smaller than 50%, then that's just supplementation, rather than the substitution, of fossil energy. I think that many people tend to overestimate change over 1 to 2 years but underestimate it over 5 to 10 years, let alone over a span of 20 years. History has proven that the intervals between each substitution are getting shorter and shorter.

For the time being, clean energies such as solar power only make up 13% of the world's energy consumption, while fossil fuels make up the remaining 87%. However, clean energy R&D is an irresistible trend. US President Obama once predicted that by 2025, renewable energy would constitute 25% of total energy usage in the world. Mr. Ban Ki-moon, Secretary General of the United Nations, also predicted that by 2030 renewables would reach 30% of total global consumption.

In the evolution of the world's energy structure, it took 100 years for coal to replace lumber as the primary energy source, while it took around 60 years for oil to take the place of coal. Today, the future of the energy structure looks to be more efficient and cleaner, with a lower or even non-existant carbon footprint. Some experts predict that renewable energies will replace fossil fuel comprehensively by 2050. I don't think it will take that long, because PV technologies are being upgraded and the cost of PV power generation is declining at a pace beyond our imagination.

Perhaps many people think that the picture I paint is illusory, and substituting clean energy for fossil fuels is just like chasing the sun in that we can see the goal, but never reach it. However, as a corporate executive working in the industry's front-line, I have first-hand experience—I see cut-throat competition every day, and I know what the market is like. I believe that this first-hand experience tells me that this substitution will not be out of reach. Jeremy Rifkin is an active advocate of clean energy, but he is still amazed by the fast development of technologies used in the relevant companies' practices.

After 2000, the solar power industry around the world achieved within just five years an annual growth rate exceeding 40%. Back in the early 1990s, global production by solar cells was less than 100MW, while production soared to 287.7MW with an annual growth rate of 20% in the late 1990s. Between 2000 and 2006, production further jumped to 2.5GW.

In 2011, global production by PV cells had reached 37.2GW, a 36% increase compared with the 27.4GW of 2010. In the first quarter of 2012, it registered a year-on-year growth of 120%. Despite anti-dumping and anti-subsidy sanctions from the EU, China's production in the first half of 2012 was still as high as 11.3GW, making up 63% of the global total.

In addition to increased production, advancement in solar technologies is particularly remarkable. Hanergy identifies thin-film technology as its future direction. Thin-film cells can be divided into three types based on the materials they use: silicon-based thin-film cells, compound thin-film solar cells, and dye-sensitized solar cells. Among these three types, only amorphous silicon (a-Si) thin film, cadmium telluride (CdTe) thin film and copper indium gallium selenide (CIS or CIGS) thin film have been commercialized.

According to the *Research and Forecasting Report on the Global and Chinese Thin-film Solar Cell Market (2013–2015)* published by OFweek's Industry Research Center, the global thin-film solar cell industry emerged in 2005, and the production of thin-film solar cells grew from 100MW to 1,981MW just 5 years later.

The report forecasted that the global production of thin-film solar cells would go up by 30% in 2013 compared with that of 2012, and by around 25% annually between 2014 and 2016. It is estimated that the global production of thin-film solar cells will reach 12.5GW, with a production value of USD 7 billion, by 2016. At the same time, the generation cost of thin-film solar cells keeps on decreasing. For example, MiaSole, a company that owns the most advanced thin-film technologies in the world, has mass-produced CIGS thin-film PV components with a conversion rate of 15.5%. By 2014, the production cost of the industry will drop to USD 0.5/W.

In multiple ways, PV companies are promptly delivering solar power to people's lives.

First, all countries are coming up with their own "rooftop program" to install solar power generation systems on the roofs or outer walls of homes, offices and public buildings, and connect the systems, and transmit power, to the grid.

Second, PV companies are partnering with property developers based on a new construction model, presenting solar power generation as a highlight of modern building design. Accordingly, solar cells are integrated into the buildings to replace traditional construction materials. In this way, they can be connected to the public power grid as well.

Third, battery modules are installed in the systems and are connected to solar power generation systems via charge controllers. In this way, the generated power can be stored for future use.

Fourth, new markets can be explored for solar power, e.g., the development of solar power lighting and solar power automobiles, etc.

Finally, thin-film solar cells are easy to carry and curl, and therefore can be used to manufacture consumer electronics and small appliances.

With all of these incentives, the springtime of the PV industry is coming.

For China, the door to the PV application market is open. China will evolve from a large PV manufacturer into a leader in PV application. Through large-scale structural adjustment and the application of better PV products here and there, China will tremendously lower the cost of power generation, thus eventually realizing grid parity. We'll see that day in the near future. In the second half of 2012, Italy achieved grid parity, and China is about to realize it in 2 to 3 years.

Once the generation cost of solar power is equal to or lower than that of traditional fossil energy, substitution won't be far off.

ENERGY INDEPENDENCE: CHINA TAKING THE LEAD

Introduction

In the era of fossil fuels, coal has been called "the food for industry," and oil "the blood of industry." To gain their "food" and "blood," Western powers manipulated geopolitics and waged world wars and energy wars.

In today's world, the US has a predominant say in worldwide energy politics, while China, the second largest importer and consumer of energy, needs to import 61% of the oil it uses. How can China get rid of the geopolitical constraints, achieve energy independence and ensure energy security while continuing to "rise peacefully"?

War is by no means an option. In the new round of energy revolution, the world will choose China, who will choose PV.

The Avoiding of An "Energy War"

On January 18, 2013, the US Department of State held a routine press conference in the Press Room in the west side of the ground floor of the Truman Building. Victoria Nuland, spokesperson of the US State Department, was at the center of this event, which was attended by scores of journalists from renowned media outlets from around the globe.

One journalist asked: "Do you have anything to say about the latest talks between International Atomic Energy Agency (IAEA) inspectors and the Iranian Government?"

Ms. Nuland answered that Iran had failed to come to an agreement with the IAEA on a "structural approach" during the dialogue held in Iran's capital, Tehran, between the 16th and 17th, and that the US was obviously "deeply disappointed" that Iran has once again missed an opportunity to cooperate with the IAEA.

Similar tug-of-wars between the US and Iran have been repeated for 30 years, with the Iran nuclear issue being one of the hottest political topics in the world. In the 1950s, Iran started developing nuclear energy with the support of the US and other Western countries. In 1980, when its ties with Iran were severed, the US accused Iran of using "peaceful utilization of nuclear power" as a cover for developing nuclear weapons and adopted "containment" policies towards Iran. The IAEA passed several resolutions on the Iran nuclear issue. In June 2010, the UN Security Council approved the most severe sanctions in history against Iran.

All of this reflects the current energy landscape in the world. On the surface, the US seems to be concerned about Iranian development of nuclear weapons. Actually, what really worries it is oil in the Middle East, and it is especially worried that an important oil-transporting channel—the Strait of Hormuz—could be controlled by Iran, its archenemy.

As stated, in the era of fossil energy, coal is called "the food for industry," and oil is called "the blood of industry." Industry cannot be developed without "food" or "blood." Nonetheless, uneven distribution and the imbalance between supply and demand of fossil energy are critical to the political order in the world. They also give rise to so-called "geopolitical" issues and have even escalated into full-blown conflict between countries, e.g., the Franco-Prussian War.

Otto Bismarck, then Prussian Chancellor, launched the Franco-Prussian War on purpose in 1870 by occupying Alsace-Lorraine. The abundant iron ore in this region together with coalmines in Ruhr enabled Germany to become the largest economy in Europe while the French economy floundered.

The Franco-Prussian War set the stage for the "geopolitical" theories of Friedrich Ratzel, a human geographer in Germany, who took land as a key incentive for conflicts and wars between countries. In 1904, Halford John Mackinder, a British geographer, proposed the "Heartland Theory," saying mineral resources were at the core of conflicts in the world. In the same era, Alfred Thayer Mahan, a US Navy captain, came up with the "sea power theory" because many straits had become crucial channels for carrying oil.

These concepts and others concerning land, resources and energy pushed colonialism through the whole world. Just as Mackinder put it in his book *The Geographical Pivot of History*: "Who rules East Europe commands the Heartland; who rules the Heartland commands the World-Island; who rules the World-Island controls the world." (More details in Case Study 1.)

The Franco-Prussian War and Geopolitics

In the middle of the 18th century, as the Industrial Revolution gathered steam, European countries lead by the UK entered the era of industrial civilization. Imperialist powers contended with each other for land and resources, dividing up rare mineral resources around the globe through violence. Consequently, even Europe was trapped in constant conflict. Taking France and Prussia as example, Alsace-Lorraine in France was endowed with large amounts of iron ore as the largest iron production zone in Europe, while the Ruhr Basin in Germany was home to the largest amount of coal in Europe. The country that took the other's resources by war would win the throne.

Then-Prussian Chancellor Bismarck infuriated France on purpose in 1870, triggering the Franco-Prussian War. French Emperor Napoleon III, Marshal MacMahon, 39 generals, 300 military officers and 86,000 soldiers were captured in the Battle of Sedan, the most disgraceful failure in the history of France. After the battle, Alsace-Lorraine, the region that had guaranteed the necessary resource supply for France's continuing economic development, was occupied by Germany. The Last Lesson, a famous novel by the French novelist Alphonse Daudet was based on these historical events.

The German occupation of the Alsace-Lorraine region provided mineral resources for the industrial development of Germany, and developed into an industrial base for it. The "marriage" between iron mines in Lorraine and coalmines in Ruhr turned Ruhr into a remarkable heavy industry region pillared by coalmining, steel, chemistry and mechanical industry. The industrial output of this region made up 40% of that of Germany's total, and its steel production grew from 140,000 tons in 1870 to 20.5 million tons in 1913, taking second place below only the U.S, after surpassing the UK. In the meantime, Germany became the largest economy in Europe while France hit a slump. The Franco-Prussian War boiled down to a fight for land, coal and iron.

"Geopolitics" was a thread running through the whole process of the second Industrial Revolution. From the Franco-Prussian War to the two World Wars, from the Iran-Iraq War to the Gulf War, all the important modern wars were triggered by conflicts over energy.

In the wake of the World War I, the UK and the US controlled almost 75% of the raw mineral materials of the world, and Germany, Italy and Japan had only 11%. When a worldwide economic crisis broke out in 1929, the world was mired in an economic downturn, with Germany, Italy and Japan revitalizing their economy by waging war for resources.

On the eve of the World War II, Hitler proclaimed his "oil economics" to the press by saying, "If I had the Ural Mountains with their incalculable store of treasures in raw materials, Siberia with its vast forests, and the Ukraine with its tremendous wheat fields, Germany would swim in plenty!" Similarly, Japan extended its invasion of China to the Asia-Pacific region at large, and declared war on the US and the UK to gain more resources—from the coal, gold and grain in China to the oil in Indonesia.

Oil was also blamed for many international frictions after World War II, with the Iran-Iraq War and the Gulf War as the most typically cited cases (more details in Case Study 2). For the first time, the Islamic countries in the Middle East realized that oil could be used as a strategic weapon. Hence by reducing production, raising prices, oil embargoes, and nationalizing oil companies, these countries panicked leaders in other countries. Thereafter, the "oil motive" can be seen to be the reason behind many wars. In the Gulf War, American government officials and even the press admitted that the soldiers were fighting for oil. At this time, phrases like "war for oil" also began to appear in many newspapers and magazines.

The Iran-Iraq War and the Gulf War

As the key bridge connecting the East and West, the Middle East has strategic geographic significance. Even more importantly, it is rich in oil. According to research, the oil reserves there make up 65% of the world's total. If oil is the "blood," of industrial development, then the Middle East can be seen as the "blood bank," the owner of which rules the world. That's why Iraq, a country lead by Saddam Hussein and built on oil, tried to rule that region.

Shatt al-Arab, at the border between Iran and Iraq, is a key channel for exporting oil from both countries. In 1980, to control Shatt al-Arab, and in turn further control the oil exports of the entire region, Saddam Hussein declared war on Iran, marking the outbreak of the Iran-Iraq war. The war ended with no significant gains and, like a military marathon, lasted for 7 years and 11 months.

On August 2, 1990, Saddam declared war on Kuwait by occupying Kuwait in one swift movement, realizing Saddam's dream of expanding Iraq's "oil territory." After the UN's warnings failed to be heeded several times, the full-scale outbreak of the Gulf War occurred on January 17, 1991. Thirty-four countries led by the US launched a massive military attack on Iraq, which was forced to withdraw.

Both wars ended but the oil-rich Middle East has never shaken off the aftershocks. Against the backdrop of increasingly scarce fossil fuel resources, the US is fixing its eyes on Iran, trying to control the whole world by controlling this key region.

Oil competition has lasted into the highly civilized 21st century. After the 9/11 terrorist attacks in 2001, the US waged the Iraq War for oil by using anti-terrorism and nuclear security as justification. In modern society, oil and war are seemingly inseparable.

These cruel wars make people reflect on how exactly human beings should go about eradicating "geopolitics" and "energy wars" once and for all.

Given that Europe was the fountainhead and the main battlefield of both world wars, Europeans were aware of the danger of "geopolitics" at a very early stage. With the Treaty of Paris establishing the European Coal and Steel Community signed in 1951, France and Germany consolidated and integrated their coal and steel industries, ending a hundred years of confrontation and contention. Such an approach inspired Italy, the Netherlands, Belgium and Luxembourg in establishing the European Coal and Steel Community, which evolved into the European Economic Community (EEC) and eventually, the European Union (EU). That is why Europe is leading the world in the third Industrial Revolution and the new energy revolution. Unfortunately, European countries only intend to remove "geopolitics" in Europe but still employ policies of this nature in other regions to varying degrees.

The US has not always been consistent on the subject of "world peace," the common aspiration of all mankind. The US has been the biggest beneficiary of geopolitics in the past two centuries. For example, since the 1970s, it has been the largest net oil importer in the world, and that defines Washington's diplomatic policies towards hotspot oil regions (such as Saudi Arabia, Iraq, Venezuela, Nigeria and the Caspian Sea region), namely to maintain order by playing "world police" while stirring up regional conflicts for its own interests.

So how can geopolitics and energy war be relegated to history? The Europeans tend to be overcautious in launching an energy revolution and the Americans have yet to demonstrate a clear course of action. Only China, who is pursuing a "peaceful rise," can take the lead in this matter.

> Solar power generation technology marks a major transformation in the history of energy and it will change the global landscape of energy and geopolitics. Sparing no effort in promoting solar power has great significance to the energy independence and security of all countries.

China's economic development in the 21st century changes supply and demand in oil. The 2008 financial crisis dragged the world economy into a slump, where China was highlighted as a crucial engine for the world economy. 2009 was the year when Saudi Arabia began to export more crude oil to China than to the US. According to the Chinese, the US is the biggest importer of oil, followed by China.

China didn't get its share of oil imports through war or geopolitical manipulation, but equal and reciprocal foreign relations, fair trade as well as large investment in oil-producing countries. To leverage oil in Africa peacefully, it has even provided high amounts of economic aid to many African countries.

In the future depicted by Jeremy Rifkin, oil will be replaced by distributed solar power, which will bring fairness and democracy to energy distribution, with political "dominance" replaced by "cooperation" and "sharing." Countries will feel free to share information and green energy technologies. Thanks to the "fairness" and "generosity" of the Sun, as long as a country owns the right technology, it can generate the energy it needs with no need to steal others' sunlight.

Developing solar power will be the most effective way to materialize Rifkin's hypothesis, and to prevent geopolitics and oil wars. With fossil fuels being replaced by new energies such as solar power, the negative impact of fossil fuels can be mitigated or even eliminated, and geopolitics and oil wars can be stopped once and for all.

At this time, China is pursuing a "peaceful rise," "scientific development," and the "Chinese Dream" of rejuvenating the nation. Meanwhile, China has gained the upper hand in developing solar power.

Farewell to the "Greenhouse Effect"

We are paying for the previous two prosperous Industrial Revolutions, whose consequences are more devastating than either global debt or the financial crisis. In last 200 years, coal, oil and natural gas have empowered industrialization, but at the same time, a vast quantity of CO_2 has been released into the atmosphere. This has caused the greenhouse effect, which may change the Earth's temperature greatly, subsequently threatening humanity's survival.

The environmental impact of fossil fuels includes the greenhouse effect, caused by the over-emission of CO_2, and acid rain derived from SO_2.

Every year, a huge amount of CO_2 is emitted through the burning of fossil fuels. Research on CO_2 published in the *Proceedings of the National Academy of Sciences* (May, 2007) revealed that global CO_2 emission in 1980 was around 5 billion tons, and following that year, emission kept rising to almost 30 billion tons in 2004. According to the IEA statistics, global CO_2 emission in 2012 increased by 1.4% to 31.6 billion tons, its highest in human history.

Currently, the CO_2 concentration in the atmosphere is growing by 1.5–1.8ppm every year. At this rate, it will reach 600ppm in 2030, with atmospheric temperatures increasing by 1.5–4.5°C over the current level. By the end of the 21st century, the amount will exceed 1,000ppm, an environmental catastrophe.

China is faced with numerous challenges in addressing this environmental crisis. For one thing, China didn't pay enough attention to environmental protection during its industrialization, which led to soaring pollution and environmental damage. For another, the lion's share of the energy China currently is consuming is made up of fossil fuels.

Let's compare the primary energy consumption structures of some key countries and regions in the world in 2012. Table 2–1 suggests that China was the largest consumer of primary energy in 2012.

The Chinese mainland and the US accounted for 21.9% and 17.7% of the total primary energy consumption in the world, respectively. The largest coal consumers were the Chinese mainland (68.5%) and South Africa (72.8%). The gas consumption of the Chinese mainland (4.7%) was only higher than that of South Africa (2.8%), but lower than the other listed countries and regions.

China's energy industry has developed into a coal-centered system with multiple energies complementary to each other. Most air pollutants (e.g., more than 70% of the SO_2 emission and 50% of the smoke dust) are from coal burning. And the chemical precursors (e.g., oxynitrides and amides) for PM 2.5 (particulate matters, a kind of air pollutant) are also related to coal burning.

Transforming this energy consumption structure and improving the environment in China are challenging and long-term missions.

The *National Plan on Acid Rain Control* by the China Research Academy of Environmental Science (CRAES) suggests that China has to limit its SO_2 emission to no more than 16.2 million tons per year to meet the critical load for the sedimentation of sulfur. However, according to the forecast, even based on a plan of slow development, SO_2 emission will still reach 27.89 million tons in 2020. If we opt for the fast growth plan, the amount would be 39.45 million tons in 2020, far beyond the target capacity of the environment.[1]

As a major consumer of fossil fuels, China receives pressure both at home and from abroad.

Internationally, meetings and instruments, such as the United Nations Framework Convention on Climate Change, the Kyoto Protocol, and the World Climate Conference in Copenhagen, suggest that countries are worried about climate change, an issue concerning the core interests of mankind and the future of the world's development. China has immense energy consumption needs, so it has to take corresponding responsibilities and obligations.

Environmental problems due to China's consumption of fossil energy have led to urgent and even very intense practical and social challenges.

Meanwhile, China is making efforts to improve the environment, and has had some success. June 17, 2013 marked the first "National Low Carbon Day"

Table 2–1 The Primary Energy Consumption Structure of the World in 2012

Country	Oil (%)	Gas (%)	Coal (%)	Nuclear (%)	Hydro (%)	Renewables (%)	Total (million TOE)
USA.	37.1	29.6	19.8	8.3	2.9	2.3	2208.8
Canada	31.7	27.6	6.7	6.6	26.2	1.3	328.8
Mexico	49.3	40.1	4.7	1.1	3.8	1.1	187.7
North America in total	**37.3**	**30.1**	**17.2**	**7.6**	**5.7**	**2.1**	**2725.4**
Brazil	45.7	9.5	4.9	1.3	34.4	4.1	274.7
South and Central America in total	**45.4**	**22.3**	**4.2**	**0.8**	**24.9**	**2.4**	**665.3**
France	33	15.6	4.6	39.2	5.4	2.2	245.4
Germany	35.8	21.7	25.4	7.2	1.5	8.3	311.7
Italy	39.5	38	10	–	5.8	6.7	162.5
Russian Federation	21.2	54	13.5	5.8	5.4	0.1	694.2
Spain	44.2	19.5	13.4	9.6	3.2	10.2	144.3
Turkey	26.4	35	26.3	–	11	1.3	119.2
Ukraine	10.5	35.6	35.6	16.3	1.9	0.1	125.3
UK	33.6	34.6	19.2	7.8	0.6	4.1	203.6
Europe and Eurasia in total	**30**	**33.3**	**17.7**	**9.1**	**6.5**	**3.4**	**2928.5**
Iran	38.3	60	0.4	0.1	1.2	<0.05	234.2
Saudi Arabia	58.4	41.6	–	–	–	–	222.2
Middle East in total	**49.3**	**48.7**	**1.3**	**0.1**	**0.6**	**0.1**	**761.5**
South Africa	21.8	2.8	72.8	2.5	0.3	–	123.3
Africa in total	**41.3**	**27.4**	**24.2**	**0.8**	**6**	**0.3**	**403.3**
Australia	37.5	18.1	39.1	–	3.2	2.1	125.7
Chinese mainland	17.7	4.7	68.5	0.8	7.1	1.2	2735.2
India	30.4	8.7	52.9	1.4	4.6	1.9	563.5
Indonesia	44.9	20.2	31.6	–	1.8	1.4	159.4
Japan	45.6	22	26	0.9	3.8	1.7	478.2
Republic of Korea	40.1	16.6	30.2	12.5	0.2	0.3	271.1
Taiwan, China	38.6	13.4	37.6	8.3	1.1	1	109.4
Thailand	44.6	39.2	13.6	–	1.7	1	117.6
Asia-Pacific in total	**27.8**	**11.3**	**52.3**	**1.6**	**5.8**	**1.3**	**4 992.2**
World in total	**33.1**	**23.9**	**29.9**	**4.5**	**6.7**	**1.9**	**12476.6**

Data source: BP Statistical Review of World Energy 2013

in China. According to Xie Zhenhua, Deputy Director of the National Development Reform Commission (NDRC), from 2006 to 2012, the energy consumption per unit of GDP decreased by 23.6%, almost equivalent to cutting 1.8 billion tons of CO_2 emission.

On the one hand, we have to control carbon emissions; on the other, we need to support economic development via adequate energy sources. Under such circumstances, renewable energies, especially solar power, the cleanest energy, are our best choice. Solar power generation requires no coal or oil burning, so we can reduce the usage of standard coal by 0.33 kilos, CO_2 emission by 1 kilo, and SO_2 emission of 0.009 kilos when generating 1 kWh of electricity.

And researchers predict that the potential installed capacity from building integrated PV (BIPV) in China will hit 1 billion kW in 2020, generating 1.25 trillion kWh of electricity every year and reducing CO_2 emissions by 1.3 billion tons.

Besides, distributed PV power generation can directly cut the energy consumption of buildings. The construction sector in China accounts for 25% of the country's energy consumption. Buildings in cities take a bigger percentage in the energy consumption mix there. For big cities, this share is as much as 30% of the total. Hence, energy conservation in buildings necessitates clean energy supply. Generally speaking, buildings mainly use electrical and thermal power for heating and boiling hot water, so there will be a huge market for renewable energies. To supply electricity, generation devices such as solar panels, distributed solar power stations and small wind turbines can be installed on public buildings, office buildings, and residences. Thin-film cells can be integrated into outer walls, glass windows, and glass curtains. Distributed solar power stations can partially or even totally meet buildings' demands for power, and the surplus power can even be sold to grid enterprises directly.

Thus, PV power generation can be the "terminator" for air pollution. China should give strategic importance to the renewable energies, particularly solar power, the most advantageous, because it will help China to energy independence and in its sustainable development.

Achieving Energy Independence

I once saw an impressive political caricature in the *New York Times*, in which a fat panda surrounded by empty barrels was on the ground, holding a barrel of oil, pouring oil desperately into its mouth.

The implication is obvious in the caricature. To Westerners, China is looking all around the world for, and then consuming, oil in dramatic quantities.

In 1993, China switched from being a net oil exporter into a net importer. In 2012, China's annual oil consumption registered 500 million tons, the second largest in the world. In the same year, China's total imported crude oil hit 270 million tons with a year-on-year growth of 6.8%, which made China the second largest oil importer. In addition, 26% of the oil it used was from foreign countries at the beginning of 21st century, while in 2012, that number was 56%. It's predicted that China will have to import 61% of the oil it burns by 2015. So we need to prepare for that rainy day.

China has already paid a huge price for this rising consumption. From 2006 to 2012, the Middle East raised the benchmark price (the Dubai crude oil spot price), a major reference for exporting oil to Asia, from USD 61.5 to USD 109.1/barrel. Correspondingly, the average price of China's imported crude oil went up from USD 60/barrel to USD 110/barrel.

The international average price for crude oil futures reached its historical high in 2012. The average price of West Texas Intermediate (WTI) crude oil was USD 94.2/barrel and that of Brent crude oil was USD 111.7/barrel. As reported in the media, the cost for China to import crude oil reached USD 110.31/barrel in 2012, and the total cost was as high as 1.38 trillion yuan that year. Crude oil has become the most important import item for China.

Also in the same year, China displaced the US in net crude oil imports. According to the US Energy Information Administration (EIA), in December 2012, the net oil import of the US dropped to 5.98 million/barrels/day, which is the bottom low since February 1992. However, according to China Customs, during the same period of time, China's net import of oil soared to 6.12 million barrels/day. By Comparing these figures, China has surpassed America as the world's largest net importer of oil. Such displacement will mean both blessings and curses for China.

Although the displacement might be temporary, figures tell us that the higher energy dependence is, the lower energy security level will be.

It's inevitable that China's oil consumption and import rate will keep growing. As international oil prices continue to rise, it has to spend more and more in foreign currency for this purpose, which will damage its international competitiveness, energy security and even national security.

China is still experiencing fast-increasing industrialization, technological adoption, urbanization and agricultural modernization, so we can hardly avoid rapidly consuming a large amount of resources in the short term. In my opinion, as long as we develop renewable energies, we can depend less on oil imports in the future. Only by doing so can we lower dependence on oil imports, and realize self-sufficiency in energy.

The new energy sources led by solar mark a major transformation in humanity's energy history:

1. This is an era when fossil energies can be substituted with solar power on a large scale, which will lead to a major transformation in the narrative of energy history, and will radically change the global energy landscape.
2. The new energy-lead industrial revolution represented by solar power is around the corner. This revolution will have more fundamental influence than the previous ones dominated by steam engines, electricity, and the IT industry. It will probably become the most significant industrial revolution since the beginning of human civilization.

To consume large amounts of energy is unavoidable in order to develop economically. Given that, we should use domestic, clean energies instead of fossil fuels. China will be able to achieve energy independence when it is independent upstream of energy development (supplying solar power, wind and thermal energies), in mid-stream (having core technologies in solar power generation and wind turbines with independent IPRs), and downstream (exploring domestic market demand).

Countries may choose different paths to energy independence, but one of their commonalities is use of renewable energies.

Let's take the "path of forests" taken by Finland. Since Finland is rich in forest, 50% of Finland's renewable energies are from lumber-based raw materials. What comes next for them is wind power, and then second-generation biomass energy. At the end of the day, the Finnish government has planned to substitute coal with renewable energies within 10 to 20 years, aiming to lower its coal use to zero.

Let's next examine the "path of wind" taken by Denmark, which is a leader in global clean energy technology. Its export of wind power equipment accounts for more than one-third of the total in the global market. Not only that, but the statistics of the Energy Agency of Denmark suggest that the total installed capacity of wind power in Denmark in 2010 climbed to 3.8 GW, 25% of the country's total power generation.

Let's also have a look at the renewable energy path favored by India. As early as 2010, the Indian government declared its intention to increase the percentage of power generated by renewable energy to 10% in 2015 from 4% today, to which it plans to add another 1% growth each year, eventually reaching 15% in 2020.

Another notable path is that of nuclear-solar power in Japan. Prior to the Fukushima Nuclear Disaster in March 2011, the Ministry of Economy, Trade and Industry (METI) of Japan shut the feed-in tariff (FIT) regime for renewable energies out. But after the accident, with the temporary shutdown of most nuclear stations, FIT was re-launched in July 2012, marking to kick-off of a cross-Japan solar power generation program.

The US also implemented a plan for developing renewable energy, but it still desires to achieve energy independence with shale gas. The reason for that choice is that America has the leading fracturing technologies. With the application of computer technology, it's possible for the drilling equipment to reach a 12-meter-wide core gas layer lying over 1,500 meters underground. What's more, 72% of the world's drilling equipment is on American soil, and this is exactly what is needed to explore shale gas opportunities.

From 2009 to 2012, the US increased its supply capacity of shale gas by 100 billion m³, or 84 million tons of oil equivalents, or about the same volume of oil the US imported from the Middle East. Combined with existing oil reserves, the US would not need to import oil from the Middle East for a stretch of up to 420 days. This is why the US is not afraid of any oil crisis that might be triggered by the Iran nuclear issue.

However, shale gas is not a renewable energy, as it will be exhausted some day, after all. Its exploitation is also likely to cause water pollution. Even if it enables the US to achieve energy independence, it will still end up being replaced in the PV revolution.

I strongly believe that the US has taken the wrong way to energy independence, perhaps because it does not understand the end of the oil age in its historical context or is reluctant to accept the changing landscape formed on the basis of a new energy revolution in the foreseeable future.

Above all, in bidding farewell to oil politics, history is bound to choose China, which in turn will naturally count on PV to achieve energy security and independence.

PV Revolution: China's Inevitable Choice

An Industrial Revolution is Calling China

The 18th National Congress of the Communist Party of China convened in the People's Hall on the morning of November 8, 2012. Just as all ordinary people in China did, I paid special attention to this conference, because here the ruling party of China elected its new leadership and made the strategic policies for the next stage of China's history. Thus the blueprint of "rising China" became clearer and more specific.

However, as the Chairman of the China New Energy Chamber of Commerce under the All-China Federation of Industry and Commerce (ACFIC), I am intrigued by the conference for a different reason from others: during the conference, the "Scientific Outlook on Development" was identified as a key guideline for the CPC and was included in its constitution. The core of the Outlook is "putting people first and achieving comprehensive, coordinated, and sustainable development." How, then, should this strategic idea be applied in the PV industry?

Based on two decades of exploration in both the new energy sector and the practices of individual companies there, together with the key message this conference has sent concerning China's destiny, I am convinced that in the new political and economic structure of the world, and in order to ride the tides of the third Industrial Revolution, the world needs China more than ever, and China needs solar power more than ever, too.

First, relying excessively on oil imports leads to potential energy insecurity, so China needs a more reliable energy strategy. In 2012, China had already become the second largest consumer of energy and the second largest economy in the world (replacing Japan). But its per capita energy consumption has only now reached the world's average, and its per capita GDP is only half of the world's average. In other words, its energy consumption per unit of GDP is twice the world's average level, 3 to 6 times that of developed countries, and even higher than that of India.

There are two key factors in energy consumption: total consumption and utilization efficiency. The total energy consumption of China is extremely high, but the utilization level remains low. To reverse this trend will take a long time, to say the least. Moreover, over time, China's economic aggregate will keep on expanding; living standards will keep improving, increasing China's total energy consumption as a result. The State Council decided that by 2015, China should consume no more than 4 billion tons of standard coal for energy. It's estimated that the annual total energy consumption of China will reach 5.5 billion tons of standard coal in 2020, and 7.5 billion tons in 2030. That is to say, the increase is unstoppable.

To address this challenge, there are only two options: first, further develop and purchase fossil fuels; second, look to alternative, and clean, energies.

The yield of fossil fuel in China falls far short of demand. In 2012, China's oil exploitation yielded around 200 million tons while it used up to 500 million tons in the same year. China imported 270 million tons of crude oil that year, namely 56% of its total consumption.

China has benefited from the global economic structure in the past three decades—it gained oil supplies from the Middle East through the safe energy channels dominated by the US and in the massive overseas market. As America develops shale gas opportunities, adjusts its strategies towards the Middle East, and no longer invests more money and resources in global energy channels, China will have to face amplified risks in its external energy supply. If we consider the disputes over sovereignty in territorial waters while developing offshore oil, gas and energy-related political factors, China will end up facing bigger challenges in securing its energy supply from abroad.

Therefore, the first option—further exploiting oil and gas—is not a safe one, exposed as it is to the dangers of the existing energy strategies and consumption structure in China. That pushes China towards the second option—looking for alternative energies.

Secondly, as the world's largest carbon emitter, China has to turn to clean energies to cut emissions. According to the *Annual Report of China's Low-Carbon Economic Development (2012)* published by the Chinese Academy of Social Sciences (CASS), despite being the world's largest carbon emitter, China is also the largest carbon emission reducer. From 2005 to 2010, China reduced its energy consumption per unit of GDP by 19.1%, the equivalent of saving 630 million tons of standard coal, or reducing its carbon emission by 1.5 billion tons.

As both the largest carbon emitter and emission-reducer, China is bound to be the most powerful force of the new energy revolution.

Now, at a time when countries around the globe are building low-carbon cities, China is also making substantial progress in this area. In 2010, the NDRC listed 5 provinces and 8 cities for the pilot "low-carbon economic and industrial development" program. With the guidance of the central government and support from the World Wildlife Fund (WWF), Shanghai started to plan the construction of the "Chongming Low-Carbon Eco-District," "Hongqiao Low-Carbon Commercial District" and "Low-Carbon Development Demonstration Zone of the New Town Surrounding the Port." Shaanxi Province promulgated a "Plan of a Main Functional Area in China" and established the "New Town of Xixian," an ecological and pastoral new town. Hubei Province, by leveraging its advantage in the research of low-carbon energy, came up with the slogan "Eco Hubei" and promoted low-carbon technology to large companies. Jiangsu Province has spared no effort in building Nanjing into a "Pilot City of Low-Carbon Growth" and Yangzhou into a "City with Low-Carbon Industries."

As the name suggests, a low-carbon city is about developing new energies, making innovations in low-carbon technologies, usings smart grid structure, and the recycling and efficient use of power.

Third, given that the third Industrial Revolution is to feature joint owner-ship, self-service and the sharing of new energies through technical means, this suggests that scattered production and decision-making will be common. As a result, the force behind this energy revolution cannot be a hegemonic country.

In retrospect, we can see that the two Industrial Revolutions in the era of fossil energies broke out because there was no way to address national problems through peaceful, commercial and technical means. However, the third Indus-trial Revolution allows hundreds of millions of people to be connected and co-exist in the social network of the energy web. Each individual on Earth is going to be the source of the energy he/she needs, the economy will enjoy truly free devel-opment, and personalized production will be possible. All such phenomena will weaken the basis needed for a single country to dominate the world.

Since all countries are egocentric to some extent, coal-abundant countries focus on using coal and oil-rich ones like to promote an oil-based economic system. The reason the US insists on achieving energy independence by shale gas is exactly that it has gigantic shale gas reserves and mature exploitation technology.

I don't see the need for China to stick to fossil energies. China has always promoted world peace. More importantly, after more than 30 years of reform and opening-up, it has to transform its economic system. Are there any better opportunities available to China than an energy revolution? Based on historical precedent, this kind of revolution brings a wave of industrial and technological revolutions in its wake. Assume that a small-scale revolution in shale gas could help America revive its manufacturing industry—what surprise would a world-wide PV revolution bring us?

China needs the third Industrial Revolution, and the revolution needs China. Other than the US, no country can outrun China in terms of economic aggregate, international influence, and the drive to transform domestic energy structure.

Rifkin expressed his opinion on China in the foreword to the Chinese edition of *The Third Industrial Revolution*:

If America was the role model of economic development in the world in the 20th century, then China probably will play this role in the 21st century.

In the following years, China has to make critical decisions on the direction for its future economic growth.

The Chinese should care about what China is like 20 years later. Will it struggle in the epilogue of the second Industrial Revolution, and keep relying on fossil energy and related technologies? Or, will it commit itself to the third Industrial Revolution, and the development of renewable energy and technology?

To China, building a society with sustainable development is an imperative.

I happen to hold the same view as Rifkin on this point.

The PV War has Broken Out

In the third Industrial Revolution, PV will rise to the center of the historical stage and this is also the stage for China. Though many companies have been involved in this PV revolution, Chinese companies are among the world's best in developing core technologies.

In the purification of polycrystalline silicon, the "Modified Siemens Process" has been widely used. However, it is energy-intensive and polluting, though it can generate silicon of high-purity. In this respect, China has already taken the lead by working out a better technology: a team led by Gao Wenxiu from the Shanghai Institute of Technical Physics (SITP) of the Chinese Academy of Sciences charted a new course by inventing a "physical method." On July 16, 2007, some samples from this process were tested by Japanese scientists who proved that their purity was as high as 5 to 6 N ("N" stands for ninth place behind the decimal point; purity must be higher than 4N), and the electricity and water consumed was only one-third and one-tenth of those in the "Modified Siemens Process," respectively.

With the fast development of thin-film technology, improvement in the conversion rate, and the mass production of flexible components, the advantage of the new process in cost and application will become more obvious.

This thin-film technology is the future direction for solar cells, and that is why I keep mentioning it. To be fair, mono-crystal silicon, poly-crystalline silicon and thin film have their own pros and cons in application. However, since thin-film cells, especially CIGS cells, can still generate electricity with dim light, and are insensitive to temperature changes (when the temperature goes up, the generating efficiency doesn't decrease too much), thin-film cells can produce a higher energy output.

China's PV development has also been driven by favorable policies. In 2007, China became the world's largest producer of solar cells, with the national production capacity exceeding 35 GW in 2012. The *12th Five-Year Plan for the Solar Photovoltaic Industry* published in late February 2012 made clear the country's intent to support backbone PV enterprises in China, and to help them grow. They hoped to reach the following targets by 2015: the leading manufacturers of polycrystalline silicon cells should increase their production capacity to more than 50,000 tons; the backbone PV enterprises to more than 10,000 tons; and the leading manufacturers of solar cells should increase their production capacity to over 5GW and backbone ones to 1GW. There should also be one PV equipment company with sales revenues of more than 100 billion yuan, 3 to 5 companies with sales revenues of over 50 billion yuan, and 3 to 4 companies with sales proceeds of 1 billion yuan.

Hanergy has unshakable faith in the PV industry. In the middle of 2013, it issued the *Global Renewable Energy Report*. According to the report estimates China, America and Japan together would make up 47% of the total PV demand in the world in 2013, and China would take the place of Germany as the largest PV market.

I think the renewable energy industry will gradually replace conventional energies globally instead of only playing a supporting role. As part of this process, countries are rushing to seize preemptive opportunities. Hence, Chinese PV companies should get themselves well prepared and stay alert.

Considering this, let's review the anti-dumping and anti-subsidy investigations carried out by the EU and US against China's PV products. They suggest again that EU members and the US, as big oil veterans, feel the need to face up to China's leading position in this industry before responding accordingly.

These investigations also demonstrate that the thin-film technology is indeed the future of the PV industry. The EU's intention underlying the punishment is, by containing the development of crystalline silicon cells, to help thin-film cells win bigger market share and to improve the related technologies, which are still in their infant stages. By doing so, they hope to secure the edge for Europe in next-generation thin-film PV generation.

Europe and America launched this battle to increase access to the thin-film market and win the high ground in thin-film cell technology. In my opinion, this is the main paradigm shift behind the new energy wars during the third Industrial Revolution. It is a war on trade and technology without any bombing and killing.

This is a war for leadership in new energy technologies. In a manner of speaking, the one who stays one step ahead in renewable energy technology gains the upper hand in competition in the energy sector.

To win this battle, Chinese PV companies can upgrade their technology and make the domestic market broader. *Several Opinions of the State Council on Promoting the Healthy Development of the Photovoltaic Industry* was issued in July 2013, which opened up numerous business opportunities in the Chinese market. According to media reports, some major state-owned companies have won orders to develop projects to independently supply PV power in regions that otherwise have no access to power, with the total value almost 30 billion yuan. The China National Machinery Industry Corporation and Aviation Industry Corporation of China were also approved to build PV power stations totalling 2.9 GW.

Government should be required to understand the future trajectory of the industry, and promote an industrial structure and a consumption model with low energy consumption, high recyclability, low emission and high sustainability.

A PV Strategy, Ready for Takeoff

Each of the previous industrial revolutions made it possible to "reshuffle" the ranks of power in the world. Those who seize opportunities can lead the world, like the UK in the first Industrial Revolution and the US in the second one.

History also tells us that new energies played an essential role in the rise of global powers in each revolution. Getting hold of new energies is a path for them to realize this dream. The UK controlled coal and the US controlled oil.

In the advent of the third Revolution, all major economies have reached a consensus on the necessity of developing the PV industry, by grasping key opportunities for economic restructuring and promoting the renewable energy industry. Some good examples are the Netherlands' efforts in biomass energy, Germany's in PV development, and Japan's nuclear power.

Based on the above consensus, I would like to contribute my own ideas.

First of all, experience tells us that the adjustment of the economic cycle is often symbolized by the rise of an emerging industry. Today, the solar power industry is bound to be that industry.

Second, the opportunity to develop solar PV is unprecedented. There will be a strategic depletion of oil, at the same time as PV technologies are becoming mature; so it's time to let PV dethrone the old king. Countries around the globe are rich in solar resources. Officials around the world demonstrate a common demand for PV by coming up with favorable policies and market stimulus for it.

Against such a backdrop, we believe that it's an imperative to make national policies for developing this emerging sector.

The Chinese government has set ambitious targets for green energy, namely increasing the share of non-fossil energies in the consumption of primary energies to 15%. That means we need to increase the supply of non-fossil energies by at least another 400 million tons of standard coal on the basis of the 277 million tons of standard coal used in 2010. That move will alleviate China's energy shortage to some extent.

According to the *Decision of the State Council on Accelerating the Fostering and Development of Strategic Emerging Industries* published in September 2010, as expected, the new energy industry was identified as a pillar that should be nurtured. Within this industry nuclear, solar, wind, and biomass power will be the first ones to be explored.

On November 8, 2012, the Scientific Outlook on Development was incorporated into the CPC Constitution as one of the guiding ideas during the 18th National Congress of the CPC. Its essence includes "putting people first" and "sustainability," which will make the nation pay more attention to renewable energies.

Leaving behind the rainy days of 2012, Chinese PV companies saw a rainbow in 2013 at last. On June 14, 2013, Premier Li Keqiang presided over an executive meeting of the State Council, discussing the ten measures for preventing and controlling air pollution, and ways to facilitate the healthy growth of the PV industry. Six specific measures were proposed in the meeting for this purpose, targeting the difficulties it was facing, such as insufficient power procurement and subsidies, and challenges in fundraising, etc.

These measures are:

1. That the guidance of plans and industrial policies on reasonable distribution within this industry should be improved, with special support given to distributed PV generation;
2. Grid enterprises should invest in, and put into operation, supporting grids for PV projects at the same pace, giving priority to developing those grids connected with PV stations, so that all the power generated there can be bought;
3. The policies on offering favorable tariffs for PV generation should be perfected. We should set up benchmark feed-in tariffs by region for PV plants, scale up the funds for renewable energies, and ensure the timely delivery of the subsidies based on the quantity of power generated by distributed PV stations;
4. Financial institutions should be encouraged to provide more money for the manufacturers of PV equipment and technologies;
5. We should support the R&D and commercialization of key materials and equipment, and put in more effort to work out standards and regulations for the industry;
6. We need to curb blind expansion of production capacities and encourage M&As among PV enterprises.

These measures defined the PV industry as an emerging one of strategic importance, rather than one of processing and trade. That is very important: it means that the "barbaric" growth period for the Chinese PV sector will be over at last. Instead, it will be put on the right track, correct the wrong philosophies behind its growth and get "reborn" by following the inherent rules of the industry, thus becoming an industry with real strategic significance. At present, solar power generation is still "inconspicuous" in the entire energy structure in China. In 2012, solar stations generated 3.5 billion kWh of electricity, only 0.07% of the total power output in China. This is because the market is mainly occupied by the other three established energy supply systems (the electric power grid, the oil supply system and the natural gas supply network) leaving no room for distributed PV. The clarification of relevant policies will help PV companies make educated decisions and accelerate the expansion of the domestic market for distributed PV.

These supporting measures and policies suggest that China has embarked on both building the "five major pillars" mentioned by Rifkin, and making solar PV the cornerstone of a new economic system.

Let PV Illuminate the "Chinese Dream"

In the near future, solar power will be applied to every aspect of people's lives, just as oil was when it emerged in the historical arena.

In the near future, every house can be a small-scale power station, with solar powered lighting and little in the way of carbon emission. Redundant power from each household can be transmitted to buyers through smart grids, in a transaction based on a smart network. Then, everyone will live in a global energy network and will be able to generate power, post supply and demand messages, and offer up their energy to others. Energies could be shared among countries and even continents, and countries during daytime can transmit surplus energy to other countries at night via such a network.

These dream scenarios are likely to come true in the future. Just as we could not have imagined a stream engine in the era of wood and bio-fuel energy, a moon landing by mankind during the coal era, or driving an oil-free car in the oil era, the impossible can eventually become reality.

Only when the new energy forms get incorporated into and alter people's lives can they be truly embraced by the spirit of the age. As the world has such a huge energy demand, only solar power can provide a scalable and effective approach to alleviate energy scarcity. So the PV dream will gradually come true.

In 2007, two graduates from the Business School of Stanford University started up D.light Design, a company manufacturing solar power devices, serving regions around the globe that lack access to a stable source of power. It creates a variety of efficient and reliable solar power lighting products whose retail prices are only USD 10/unit. It ambitiously aims to phase out kerosene lamps globally.

Donn Tice, the CEO of the company, was once mentored by C. K. Prahalad, the late renowned Management Professor at Michigan University. According to Professor Prahalad's book *The Fortune at the Bottom of the Pyramid*, companies can gain profits from designing products and services for the 3 billion people who spend USD 2.50 or even less for daily needs and live at the bottom of the wealth pyramid, and such good deeds will also serve to benefit mankind.

The success of Tice's company has verified Prahalad's theory. A major advantage of its products is that, almost all the electric energy is converted into light with little waste in the form of thermal energy. Such lighting products give us the best insight into the great power unleashed from technology integration.

With the integration of new energies and new technologies, there are going to be more and more such designs and products, affecting all aspects of people's lives.

> The only way to remove the "energy bottleneck" and to break the curse of the "Achilles' heel of energy" is to start a new round of energy revolution. To be specific, it should be a solar PV revolution, in which the strategy of renewable energy substitution will be implemented. And this is crucial for sustainable economic growth in China.

China is also looking forward to the changes that will be brought about by solar power. Without any exaggeration, the "PV Dream" is the carrier of the "Chinese Dream."

On November 29, 2012, the newly elected Chairman of the CPC Central Committee, Xi Jinping, gave his interpretation of the Chinese Dream when visiting the "Road to Revival" exhibition in the National Museum of China. He said: "Today, everyone is talking about the nature of the Chinese Dream. I believe that the great revival of the Chinese nation is the greatest Chinese dream." He also pointed out that the goal of building a well-off society in all respects will have been attained by the centennial of the CPC; and that the goal of building China into a rich, strong, democratic, civilized, harmonious and modern socialist country will have been attained alongside the realization of the great dream of reviving the Chinese nation by the centennial of the PRC also.

The revival of China requires sustainable economic development. Only when it becomes an economic and innovative power can China gain a solid physical and technical foundation for the great rejuvenation. At this point, only the PV revolution can ensure energy security and enable the necessary sustainable economic growth. As an executive in the solar power sector, I can draw the conclusion that only the achievement of the "PV Dream" can provide the basis for the aforementioned "Chinese Dream."

PV Revolution to Lead the World

Can technology be invented which uses solar power directly, as plants use sunlight to create energy? The answer is yes—and it is the all-in-one solar PV technology that can make it happen. At the center of the competition in solar power generation is a competition among core technologies. Those who own the core technologies can be the masters of this energy. With the policies made by its government, hopefully China will make the first moves in this competition.

A *"Quasi-Chlorophyll Revolution"*

People's primary expectation on solar power can be summarized in one sentence: to be able to use solar power in the manner of chlorophyll.

Nature, after all, does have a system to capture sunlight. From the birth of the planet's first plant life to the present day, the system has functioned for more than 3 billion years. And it is called photosynthesis.

Research on photosynthesis has occurred over 200 years and has proven quite fruitful. Ernst Walter Mayr, a German-born scientist, first found that plants had the ability to transform solar power into chemical power in 1845. Julius von Sachs, another German scientist, found that starch grains could be generated from photosynthesis in 1864. German scientist Theodor Wilhelm Engelmann discovered that chloroplasts generate oxygen in 1880. German chemist Richard Martin Willstätter discovered that chlorophyll, which is abundant in green plants, acts as a light-harvesting complex. Chlorophyll is a hollow sphere with its typical structure being one of regular icosahedral symmetry, and is filled with pigment molecules for absorbing and transferring sunlight.

The fantastic photosynthetic reaction of chlorophyll inspires people. For the benefit of our economic activities, industrial production and daily life, can we invent artificial chlorophyll that can help us directly leverage solar power?

As early as the 19th century, man discovered the photoelectric effect when sunlight interacts with material. Carrying out further research on this effect, scientists discovered the photovoltaic effect and then attempted to make solar cells with metal-semiconductor junctions. However, the technology necessary for solar cells was not invented until 1950s, with better understanding of the physical nature of semiconductors and the development of processing techniques.

In 1954, the first mono-crystalline silicon solar cell of practical value was invented by Daryl Chapin, Calvin Souther Fuller and Gerald Pearson at Bell Labs. They found if a small of amount of impurity were mixed into silicon, the strips of silicon when placed in sunlight turn this light into electrical currents. They also found that the sunlight–electricity conversion rate was around 10%.

Because the photoelectric materials in solar cells can capture solar power for photoelectric conversion, solar power could be said to be captured in an authentic "quasi-chlorophyll" way, thanks to this development of science.

Entering the 21st century, researchers made a new breakthrough in chlorophyll study by finding that some plants have pigments absorbing long-wave rays to respond to low-light environments. In 2010, researchers extracted such chlorophyll by chance from a colonial cyanobacterium in Shark Bay, Western Australia. Dubbed chlorophyll f, it has the ability to absorb red light and infrared rays with a wave range of 0.7mm–0.8mm (the wavelength of infrared rays is between 0.77mm and 1,000mm).

Since then, scientists have tried to develop cells based on photosynthesis. For instance, they have extracted chlorophyll from plants and put it in man-made films. With a chlorophyll-centered light-capturing system, the photoreaction center and 10 elements (such as manganese, iron and magnesium), light is converted into electric current via a photosynthetic system that is sophisticated and exquisite, and then turned into chemical energy in a stable state.

Today, cells that simulate photosynthesis have been invented, and called dye-sensitized solar cells. They adopt sensitized synthetic dyes instead of chlorophyll from plants. On November 19, 2011, the first new dye-sensitized solar cell project in China went into operation in northern Jiaozhou Bay, in Qingdao's High-Tech District.

Photosynthesis is one of the greatest chemical reactions in nature—and solar power generation through solar cells is also a terrific discovery by mankind, as it radically changes the way that people utilize energy.

Solar power generation is a direct utilization of solar energy that transforms the way people use energy. Someday, people will be able to directly use the inexhaustible power of solar energy like plants do today. Currently, the thin films can turn 15%–20% of the solar power they receive into electricity with no pollution and almost zero emission. This will be the ultimate form of energy utilization for mankind!

Therefore, the PV power generation, which can be called a "quasi-chlorophyll" revolution, will be one of the greatest energy revolutions in world history. By launching this revolution, humanity will regain the use of energy's original form.

Global Attitudes

The "quasi-chlorophyll revolution" is unprecedented, because it not only changes how energy is utilized, but also removes the constraints imposed by the interests of traditional energy sectors.

Internationally speaking, this constraint comes from America's oil hegemony and oil-based geopolitics. Europe and America carried out the investigations on China's PV products for two reasons: to consolidate their technical and patent barriers for new energies (e.g., the US has controlled the technologies and patents in crystalline-silicon solar power, and Germany controls those in precision machinery in the field of solar power); and to contain the development of new energy in China.

China is rich in coal. So it can continue to depend on coal as a major energy source for a number of years. However, both the *Kyoto Protocol* and the global greenhouse effect force us to follow a sustainable path in developing new energies. Unfortunately, technologies and patents related to new energies are in the hands of other countries. The only way for China to blaze a trail is to take the lead in new energy and high technologies.

Domestically speaking, we must overcome the constraint that comes from conventional energy interest groups. For example, in the US, political contentions in Congress, i.e., the oil and military lobbyists represented by the Republicans vs. the semi-conductor tech lobbyists from Silicon Valley represented by the Democrats, set obstacles on new energies' development path.

To show his confidence in energy independence and a green economy, President Obama changed the traditional tone of the 2009 Presidential Inauguration Ceremony from red to green. At that time, President Obama emphasized that future automobiles and factories in the US will be "driven by solar power or wind power" and that he believed renewable energies have both huge potential and a bright future. However, during the 20-minute inauguration speech for his second term on January 21, 2013, he only spent one minute on climate issues, and just mentioned in a general manner that the path promoting renewable energies is "long and difficult."

Behind such subtle changes is the fact that a shale gas revolution has been put on the agenda in the US, though an MIT announcement claimed that shale gas is only an "interim" plan for America as it moves towards a low-carbon future. Exaggerating its value will probably undermine the development of carbon capture and storage.

Europe meanwhile, lacking as it does a sufficient supply of conventional fossil fuels, is either contained by Russia or intimidated by the US, and hence has precious little affection for these energy forms. European oil giants (e.g., BP, Royal Dutch Shell, Total, etc.) have worldwide market networks so that European countries are not subject to the desire of traditional energy giants for profit.

However, because Europe is far from politically united, it can't easily consolidate these nation-based power companies and energy networks in an effective manner.

On the other hand, China has a great political edge in this regard. First of all, due to environmental pressures and the demand for economic growth, China is more willing to use new energies than any other country. In the last two decades, China has been promoting the need to turn to green energies. Second, Chinese leaders attach great importance to the third Industrial Revolution and strategic planning for the advent of these new energies. Third of all, almost all the conventional fossil energy companies in China are owned by the state, so when the government makes its energy policies, it has no need to worry about great pressure from groups with vested interests, unlike the American government.

On September 12, 2012, the NEA published the *12th Five-Year Plan for Solar Power Development*, requiring that by 2015 the installed capacity of solar power be above 21GW and that distributed PV power generation exceed 10GW.

On September 14, 2012, another NEA document titled *Notice on Application of Large-scale Usage Demonstration Area of Distributed Solar PV Power Generation* was sent out to all the solar power companies in China. According to the notice, a "distributed power generation program with an installed capacity of 15GW in the first phase" was launched, and energy departments of all provinces and cities were required to report their implementation plan before October 15 of the same year. This showed the sincere willingness of China to support PV companies and develop new energies.

And if we compare the two documents, we see that the latter set higher standards than the former, by increasing the installed capacity requirement of distributed PV power generation to 15GW, 50% higher than the number in the previous plan.

It is said that the NEA is working on the 30-year and 50-year plans for the PV industry. The total installed capacity of China will be increased to 35GW by 2015 and 100GW by 2020. From 2020 to 2050, the installed PV capacity will keep on growing by 30GW each year.

China's Strengths

The PV revolution lies directly in China's path, for three main reasons:

1. China is technologically prepared. China has progressed quickly to master leading PV technologies. For instance, Hanergy started its R&D on thin-film cells several years ago.

> On June 15, 2013, the Thin-film Forum and the Inauguration Ceremony of the Thin-film Special Committee of the China New Energy Chamber of Commerce under the All-China Federation of Industry and Commerce (ACFIC) kicked off in Beijing. Two hundred people, including insiders and experts from the PV industry and journalists from around the globe, attended the forum. In my speech, I stressed that core competition in new energies represented by solar power was different from that in conventional energy sectors: It was a competition *between* core technologies rather than one for resources. The one who own core technologies get the edge.

The *12th Five-Year Plan for Solar Power Development* has highlighted the development of crystalline silicon cells and thin-film cells, as well as research on new cell technologies. China will endeavor to improve the industrial conversion rate of mono-crystalline silicon cells, polycrystalline silicon cells and amorphous thin-film cells to 21%, 19% and 12% respectively by 2015. It will also industrialize the production of CdTe and CIGS thin-film cells. China will fulfill parity for grid-connected PV systems on the consumer side, and ensure that the on-grid price of PV electricity in the public grid is lower than 0.8 yuan/kWh.

2. The Chinese government spares no effort in boosting investment in clean energies. According to statistics from Bloomberg New Energy Finance (July 2013), in Q2 of 2013 China invested 13.8 billion USD in clean energy, with a month-on-month growth of 63%. Its total investment was also the largest in the world. In the same period (2013), the global investment in clean energy registered 53.1 billion USD, increasing by 22 % compared with Q1, which was extraordinarily higher than the 43.6 billion USD of Q1, but still less than the 63.1 billion USD of Q2, 2012. Moreover, China's investment in clean energy in Q2 accounted for 25.9% of the world's total in this category.

3. China has a vast market. Based on the NEA plan, in the 12th five-year plan period China is going to diversify and apply solar power generation on a large scale. First, China will build and integrate solar power stations into the grid with a total installed capacity of 5 million kilowatts in western regions, including Tibet, Inner Mongolia, Gansu, Ningxia, Qinghai, Xinjiang and Yunnan,

which are all rich in solar power. Second, China will strive to promote distributed solar power generation on urban rooftops. Economic development zones and industrial parks with huge rooftops are the focus and users are encouraged to use and generate power by themselves. Third, in 100 new energy demonstration cities and 200 new green energy demonstration counties, China is promoting the application of solar power generation and carrying out research on how to coordinate the development of PV power generation and smart micro power grid systems, in order to build technical and managerial support systems for distributed solar power generation. By the end of 2013, the accumulated solar power capacity of China will reach 10 million kW.

PV Revolution to Lead the World

The ultimate energy substitution is unveiling itself. Renewable energy-based distributed power generation will bring a new revolution to the world with sunlight as the hero in the process.

We have discussed the inevitability of solar power being the star in the ultimate substitution. Now, let's imagine the future with this in mind.

Man will enter the "post-oil era," which lies between the oil era and the PV era. At this stage, technological competition will center around core technologies.

We are sure that the future belongs to renewable energies, and given their characteristics, it is only by leveraging the modern technologies of equipment manufacturing can we transform geothermal energy, wind power and sunlight into controllable and usable energy. That's why we have to proactively invest in R&D to take this opportunity preemptively.

Development from R&D to industrialization takes time. Many countries will do their best to gain the technical edge. Consequently, it's possible for the countries obsessed with the current energy hegemony to wage trade wars to attack their competitors. In the next decade, trade wars will only intensify.

We have to watch out, as developed countries with existing technical and capital advantages are more likely to become the main suppliers of new equipment and materials. From the technical vantage point, they are more likely to become the main producers and controllers of devices of higher value, the major new equipment and new materials from around the world. As a result, their real economy will be further strengthened.

In this, the oil era, "technology conspiracy theories" are nothing new. For instance, America has lots of strengths and dozens of years of experience in the semi-conductor industry—so why did it choose crystalline-silicon instead of thin film for developing solar power? Because a group of leaders in semi-conductor

R&D from Silicon Valley advocated the choice. These people have a wealth of experience in large-scale integrated circuits, flat-panel displays and computer hard drives. Though technically speaking panels and semi-conductors are products of different form, they employ many of the same techniques and are made by the same production tools. What matters most is that the Americans control the critical technologies in both energy and semi-conductors. The first semi-conductor revolution brought us computers and consumer electronics. In the second semi-conductor revolution, semi-conductors are to be turned into sources of energy.

Prior to the hi-tech revolution, semi-conductor chips and circuit boards in the US emerged from the power grids and power transmission technologies of the energy industry, and the processer chipsets of semi-conductors brought about the mass production of radios, TV sets, computers and other consumer electronics.

By exaggerating the advantages of crystalline silicon, the US expects other countries to join it in developing this technology. In truth, the US shares only the lowest end of the industrial chain with minimum profit to others, while it keeps vital technologies in US hands.

We can discern the strategy of the US: connect the development of new energies to its present control over key technologies. It steers the world's technology development with this union to guarantee its international competitiveness.

Thus, if a country wishes to enhance its international competitiveness, it has to break the dominance of the US and open up a brand-new technological path. Similarly, the only way to take preemptive opportunity in the "PV era" is to make breakthroughs in technology and be among the first to enter this era.

The key for us in taking the lead is solar power technology. Today, China has already taken one step ahead of the rest, and with advanced technologies, a vast market and government strategies, China is bound to be the winner in the PV revolution.

The development of the new energy industry brings China a once-in-a-blue-moon opportunity:

1. Upgrading our solar power sector from a crystalline-based one to a thin-film-based one, maintaining of China's edge in the thin-film field and expediting the development of the thin-film solar power industry have strategic significance to the "steady growth, structureal adjustment, and industrial upgrading" of China.

2. The development pace of the new energy industry exceeds our expectations. For China, this is an unprecedented chance. If we grasp it, then China will achieve energy independence and energy security.

3. This new energy revolution is rooted in the manufacturing industry, in which China has traditional advantages, so it is absolutely possible for the Chinese to rule the game!

4. In comparison to the US, Europe and Japan, China does have unique strengths in developing the solar power industry: funds, technology and its market are the three indispensable elements. Hijacked by its interest groups, the US focuses on shale gas in exploring new energy. Europe has both technologies and market but lacks funds. Japan owns funds and technology but lacks enough of a market. Only China boasts all of them.

If China takes the lead in entering the PV era, we can bid farewell to the traditional pattern of industrial and economic development that has existed for more than 300 years. We will build a brand new industrial model and system—the PV industrial system, which will be of higher efficiency, will see faster development and will be more easily shared.

At that point, everyone will be able to generate clean energy. People at home, by leveraging the energy web, will be able to do personalized manufacturing with smart equipment and make diversified products in small quantities. This will change the simple and repetitive production model, blend the boundary between the manufacturing industry and IT and other smart industries, and will marry invention with manufacturing, tremendously shortening the start-up time for commercialization.

When China enters the PV era, we will get inexhaustible energy with almost zero fees. This is because on an energy network, the extra energy generated by each micro power station can be transmitted and traded, cross-continentally.

When China enters the PV era, we will be able to get rid of "polluting," "limited," "unsustainable" and outdated models of economic development. Meanwhile, we will say goodbye to world-haunting environmental problems like smog and the greenhouse effect.

When China enters the PV era, we will see an energy supply system that is completely different. Comprehensive and sustainable development is no longer a dream, and China will be able to be the owner of its own growth. With the change in comparative advantages and industrial structures in different countries, the economic and geographical landscape in the world may be reshaped as well. The division of labor in the global economy based on the second Industrial Revolution will be altered, and China will gain the initiative in the new round of the industrial division of labor and wealth.

When China enters the PV era, we will see no more of the geopolitical impact of fossil fuels. Thanks to the "fairness" of the sun, international conflicts will no longer take the form of energy wars, but of technological competition. Billions of people can share renewable energy over a green power network that reaches across continents. This will lay the foundation for the world economy and the harmonious development of human society.

OPPORTUNITIES AND CHALLENGES: THE NEW LANDSCAPE OF THE PV INDUSTRY

Introduction

After missing two Industrial Revolutions and two energy revolutions, China is finally ready for the third Industrial Revolution. This time, it will catch up. In this new, PV-based energy revolution, China and the developed world stand together at the same starting line.

The EU anti-dumping and anti-subsidies investigations forced us to learn from our previous crystalline silicon model, which is technically controlled by other countries and is not available to broader markets despite large production capacity. In the wake of this, we must identify the direction of our industrial transformation: thin-film and flexibility should be the strategic choice for China's solar industry.

The world's PV industry will choose a China whose PV industry chooses thin-film.

From a Requirement to a Natural Occurrence

The documentary series *Planet Earth* tells the story of an evolving Earth, and it shows the necessity of replacing fossil fuels with a new energy. So then, what is the inevitable choice for realizing the new energy revolution? It is neither wind power nor hydropower; neither nuclear energy nor biomass energy; and it is not tidal power or shale gas—because all of these energies are either unsustainable, cause new environmental problems, or involve technically insurmountable obstacles. The only feasible solution is solar power, which is inexhaustible, is developing technologically and falling in terms of cost faster than most can imagine, can replace traditional fossil fuels, and has the potential to break through the "glass ceiling" limitations encountered by other new energies. This is why the PV revolution is not an option but a necessity.

Beginning with the Beautiful Earth

Without the Earth mankind would have nothing, so for this chapter, we begin by looking at the documentary series *Planet Earth*.

It took the BBC five years, the most advanced photographic equipment, and its best photographers and pilots to finish this beautiful TV series. The documentary captured an Earth full of life from an amazing perspective.

It tells the story of an Earth whose appearance hasn't changed for several thousands of years. You will probably be blown away by every frame. However, the Earth today is totally different, because human activities have destroyed the living environments of other creatures on Earth. Every year, the ice in the Arctic Ocean thaws earlier and earlier, while the drought in Africa gets worse and worse.

Alastair Fothergill, the director of the series, said: "I hoped to capture the most beautiful side of the Earth in the series, but I'm afraid that my children might never see such real beauty with their own eyes, let alone my grandchildren."

Alastair was disappointed by the overexploitation of nature, chiefly via the development and use of oil, coal and other fossil fuels, which is causing irreversible damage to the Earth. The greenhouse gases humanity emits lead to global warming. And once the hole in the ozone layer formed, Earth began to be exposed to more powerful and harmful ultraviolet rays.

Faced with the prospect of global warming, officials are arguing over how to distribute the carbon emission quotas between different countries. Some support putting more money and time into shale gas development, and others are advocating relatively clean energies, e.g., nuclear and methane. In my view, taking extra time to discuss these issues is irresponsible. It is clear that using clean and sustainable energies to replace traditional ones is the right way forward. Fossil fuels have caused this crisis, and we don't have much time left before the end of the world as we know it, so we have to join the new energy revolution as soon as possible and start making breakthroughs there.

As of now, we are still in an extended oil era, or perhaps in a transition period from the era of oil to one of the new energies. However, depleted fossil fuels and pronounced environmental issues are forcing humanity to launch a new energy revolution.

Since the beginning of the 21st century, oil has become more expensive. The global economic crisis, which happened in 2008 and is still haunting the world, is thought to be closely linked with rising oil prices. The crisis is accelerating the demise of the era of fossil fuels. For one thing, oil price hikes result in synchronized price hikes for all economic factors. For another, continuous and huge oil price increases directly lead to the reallocation of global wealth. To China, a country highly dependant on oil imports, the consequence is lower total profits, which eventually will impede sustained, healthy and fast economic development.

We mentioned before that every historic leap in the progress of human society is rooted in an energy revolution, and that every historic recession is intertwined with an energy crisis. If we really want to avoid recessions and make progress, we have to start a new revolution.

Leading the New Energy Revolution

It is known that energy, food, and the environment are the three challenges humanity will face in the 21st century, and the energy issue ranks top of the three. In recent years, to embrace the transition from the post-fossil-energy era to the new energy era, all countries are proactively developing new energy technologies, and subsidizing or supporting new energies with preferential policies.

There is no international consensus on the definition of "new" energies. In China, they are considered wind power, solar power, biomass energy, hydro energy, nuclear energy, geothermal energy, unconventional gases (shale gas), etc. Many nations are promoting several new energies simultaneously. In the last decade, China has made some achievements and progress in wind, solar, biomass, nuclear, and geothermal energy.

However, years of experience tell us that all new energy developments cannot be called "energy revolutions." Just as we set priorities for economic growth, we also need to prioritize some energy substitutions by identifying a particular energy form as a primary strategic direction, and making others complementary to it.

Which new energy form is our best option? It should meet the three criteria we have come up with: First, it has to be inexhaustible, or at least recyclable in case of depletion. Second, it has to be pollution-free (including the energy per se, the manufacturing of related equipment, the engineering processes, and the consumption process of the energy). Third, the related technologies have to be something that mankind has already mastered, and that have plenty of scope for improvement.

There is no other energy source that can meet all the criteria above other than solar power. And experience has proved that all the other energy forms have too many serious flaws and cannot trigger a "revolution."

In Chapter One, we discussed how two new energies—wind power and hydropower—were not qualified leaders in the "new energy revolution."

Now let's talk about why biomass energy, tidal energy, nuclear energy and geothermal energy cannot fill that role either. Shale gas is more complicated, so it will be discussed last.

Biomass energy has hit its glass-ceiling, sooner even than wind.

Several years ago, many people believed that biomass energy had a lot of potential, so a great number of bio-ethanol projects were established. However, today almost all of those projects have sunk due to shortage of raw materials. China has a large population with relatively small amounts of arable land, so there is insufficient land available for bio-energy production.

Biomass energy production not only requires land but also crops. Research shows that producing every ton of bio-ethanol requires 3.3 tons of maize, 7 tons of cassava, 10 tons of sweet potatoes, or 15–16 tons of sweet sorghum.

The UN once predicted that by the middle of 21st century, global food and fuel demands would double those of the day. The organization also warned that developing biomass energy would lead to fewer food supply sources and higher food prices, especially in some poor regions like Africa.

A staggeringly large amount of land for growing bio-crops would be needed to produce enough bioenergy to replace gasoline. Every year, America consumes 530 billion liters of gasoline and 151.4 billion liters of diesel. Even if America produced bio-diesel with all its bean crops, only 6% of the total diesel demand could be met. In recent years, around 40% of the maize in the US is converted into ethanol annually, replacing only 9% of gasoline consumption. Moreover, energy is consumed in sowing, harvesting, transporting and converting maize. So bio-ethanol reduces very little carbon emission. Seeing that bio-fuels, made at such great cost, only reduced energy-related emissions by 9%, the OECD decided to stop subsidizing the bio-fuel industry.

Geothermal energy is equally unlikely to lead the new revolution. It is a kind of natural thermal energy derived from the crust of the Earth. As a form of heat, triggering volcano eruptions and earthquakes, it originates in the internal magma of the Earth. The simplest and most cost-effective way to acquire it is to directly extract the energy and generate power from it. And China has the largest proportion of geothermal energy in use in the world today.

Geothermal energy is also a renewable energy. However, it is left out in many recent energy research monographs. Even America has not included funding for geothermal energy research in the government's annual fiscal budget. Despite its status as one of the "mature technologies," geothermal energy hasn't attracted much attention from American companies compared with the highly subsidized wind power and solar power.

As in the fledgling periods of other renewables, the development of geothermal energy is restricted by technological issues and a lack of funds. The geothermal industry is capital-intensive and finds it difficult to attract investment because it takes so long to get returns following investment. We don't yet have sound technical specifications or standards for this energy form.

Besides, it has innate defects. Although the size of reserves is large, it's difficult to develop due to geological constraints. Directly discharging geothermal water may pollute surface water and soil, and the over-pumping of the water will result in ground subsidence. Therefore, geothermal energy can hardly be expected to lead the new energy revolution.

And nuclear energy is not counted as a typical new energy. Only in a broader sense is it included in the same category as wind and solar power.

Since March 11, 2011, when an earthquake destroyed the Fukushima Nuclear Plant in Japan, nuclear energy has fallen on hard times throughout the world. Many countries, including Germany, declared their intent to shut down their nuclear power plants. China also suspended new nuclear projects before coming up with new security standards. Although only a handful of nuclear power plants (including the infamous Chernobyl Nuclear Power Plant) have ever been hit by such disasters, accidents bring about serious consequences. Currently, new nuclear projects, especially onshore ones, are highly controversial, with the prevailing attitude being the "not-in-my-backyard" mindset.

Nuclear energy falls in the category of new energies because it is free from pollution and greenhouse gases. However, it is not renewable, because someday the resources it requires will be exhausted, even sooner than fossil fuels. Nuclear energy derives from uranium, whose reserves are limited. Reports show that uranium development can only last 70 years based on the current quantity of demand and exploitation in the world. And if we were to substitute all fossil energies with nuclear energy, the world's uranium stores would be depleted within 4 years.

What's more, it takes sophisticated technology and a large time investment to set up a nuclear plant, whose demolition cost is then higher than that of construction. Terminating and removing a large nuclear power station costs USD 200–500 million, and the radioactive waste will remain for thousands of years without any satisfactory disposal solution yet. Hence, nuclear energy cannot be a leader in the revolution.

Another popular renewable energy is marine power. This refers to the renewable energy received, stored and released by the ocean in all its physical processes, and the energy is in the form of waves, tides, temperature differences, salinity gradient, and current. Even with scores of years of technological development, marine power has not been utilized because of poor yields, high cost and unsolved technical problems. In addition, large projects may jeopardize natural flows, tides and bio-systems. This is why it develops slowly.

For instance, tidal power is used relatively widely. However, its development is constrained by geographical factors despite long coastal lines. First of all, tides must be high enough, as high as several meters at least. Second, proper coastal terrain is needed for storing plenty of ocean water and for construction purposes.

There are two types of coastal terrain in China—plains and bays. For example, most coasts north of Hangzhou Bay are plain coast, formed by sand and sludge with straight coastal lines and flat terrain, having small tidal ranges and lacking the required high quality dam sites for power generation. The coasts in the south are mostly bay coasts, with precipitous terrain, twisted coastal lines, steep slopes and deep water. Thus, there are significant tidal ranges in the bays and coasts, and also excellent dam sites. However, the coasts of both Zhejiang and Fujian Provinces are silt coast, in which the abundant tidal power can't be properly developed without an efficient way to remove the sludge from the reservoirs. Because of the above factors, marine power cannot lead the ultimate substitution of old energy.

All in all, none among the aforementioned power forms—wind power, hydropower, biomass energy and tidal power—is right for leading the new energy revolution.

PV is an Inevitable Choice

After analyzing the characteristics and trends of wind power, biomass energy, geothermal energy and marine power, we arrive at the conclusion that they can only supplement existing energies rather than replace them. This is not because of their small share in the energy mix, but because we see little possibility for them to be applied further or to substitute for the old system on a large scale, based on our objective analysis of the pros and cons as well as their potential to develop. Similarly, we cannot tell the future ranking of these energies on the basis of their current shares in the energy consumption pie. For example, wind power and nuclear energy represent bigger proportions of the new energy mix, but we cannot jump to the conclusion that "the future of new energy lies in them."

After various comparisons, we find that solar power is the rising star in the new energy revolution, and PV specifically will be the one to lead it.

Of course, solar power is not perfect, but the negative impact of its shortcomings can be more or less avoided or cushioned with the right technologies.

Solar plants can be accused of taking up too much land. For instance, a 1,000,000 kW PV station or solar-thermal station occupies an area of 1–3 km². However, the mainstream method in PV power generation is BIPV, which perfectly incorporates solar components into buildings. Thus, solar power generation can still work even on limited urbanized land areas. Moreover, large and medium-sized power stations can be installed in deserts, taking up no arable land.

Besides, despite high initial cost, solar power is highly adaptive, and almost free from both maintenance costs and pollution.

> The sun is the origin of all other energies. The coal-fired power, nuclear power, and oil we use today are indirect uses of solar power, and are very inefficient. For example, by utilizing traditional energies, people only make use of 1%–2% of available solar power, and the only way to get this amount of power is by burning fuel, with 98%–99% of the rest being emitted and thus polluting the air. In contrast, solar power generation is the direct use of solar energy, and will radically change how we use energy.

At this point, the photoelectric conversion rate of thin-film cells has already reached 10%–20%, over 10 times that for traditional energies. It will climb to dozens of times higher levels in the not-so-distant future. For now, it's time for solar power to replace traditional energies in a massive way. Perhaps someday people won't need to worry about energy shortages, but instead about energy surpluses!

Solar power is also the fastest-growing form of clean energy. Although it only meets a small portion of energy demand today, it has the potential to meet the world's entire demand with just 1% of the Earth's land area necessary for installing photoelectric equipment. Within around 20 years, the amount of electricity generated by solar power on land could be equal to that available from the total reserves of fossil energy.

President Obama said that renewables will make up 25% of power demand in America by 2025. UN Secretary General Ban Ki-moon predicted that 30% of traditional energy will be replaced by new forms by 2030. They are optimistic because it's much easier to build solar power facilities than to set up facilities for hydropower production or other energies. For example, it took 8 years, from start to finish, to complete the construction of the Jin'anqiao Hydropower Station, which boasts an installed capacity of 3 million kW, near the Jinshajiang River. It takes only one year to complete the construction of a solar power station with the same installed capacity. It is in part because building a solar power station is so easy that we believe solar power is the future.

Nowadays, the US, Japan and the Middle East invest hugely in R&D for solar power. It is fair to say that solar power generation is a remarkable transformation in the energy history of humanity and it is going to change the energy landscape forever. And the foundation of the transformation is technology. Since solar power is shared by everyone and cannot be monopolized by anyone, the one who owns core technologies will ultimately win out.

China, America and Europe have their own advantages in developing solar technologies. China takes the lead in silicon-based thin-film technology, but the US and EU are doing better in CIGS thin-film technology. Chinese companies can, however, get access to core technologies by acquiring US and EU companies, including their IPs.

Is Shale Gas a Flawed Proposition?

George P. Mitchell, an energy tycoon born in Texas, started the "shale gas revolution" on his own, and thanks to this he was presented with the Lifetime Achievement Award by the Gas Technology Institute (GTI) and ranked on *Forbes's* list of Billionaires.

After all, shale gas is still traditional fossil energy. And behind this crazy "revolution," manipulators like investors at the Chicago Board of Trade (CBOT), and on Wall Street, and politicians from Washington's Capitol Hill are everywhere, making the "revolution" in truth more like a scam. From the beginning of the "shale gas revolution" to the collapse of natural gas prices in 2012, two-thirds of American drilling companies were trapped and lost everything.

Therefore, we have to see the "shale gas revolution" as merely a case of falling for a trick.

An Old Man and His "Revolution"

Following the death of the 94-year-old George Mitchell on July 27, 2013, a great number of articles came out that memorialized the old Texan. He is the father of shale gas technologies, and many people believe that he alone is responsible for starting the "shale gas revolution."

Several years ago, many Americans had no idea what shale gas was, but today, numerous shale gas fields have been set up and all of a sudden shale gas plays an important part in the local economy and landscape: "If you pass a gas field by car at night, you will see shimmering derricks like Christmas trees."[1]

In 2010, GTI presented Mitchell the Lifetime Achievement Award. He improved hydraulic fracturing or "fracking," which ushered America into an era of booming shale gas development, and helped it achieve real energy independence within a matter of years. People gave credit to Mitchell for his contribution to the economy, and Mitchell himself ranked 239th among the 400 richest people worldwide according to *Forbes*.

The rise of shale gas is deemed a revolution. However, it still stands on the same team as natural gas and traditional fossil energies—it was GTI who presented Mitchell the award, and it's traditional oil companies who invested in shale gas.

Hence whether shale gas is a new energy is still in a question, and equating the "shale gas revolution" with a "new energy revolution" is not appropriate. Shale gas is an unconventional natural gas, and does not meet the criterion of "being renewable and pollution-free" defined by the United Nations Environment Program (UNEP). In addition, it's more costly to exploit shale gas than natural gas, and the exploitation consumes plenty of water and even pollutes groundwater.

However, shale gas prevailed in the US and quickly went viral around the globe, including in China.

It is reasonable for people to be concerned that the alleged "shale gas revolution" may weaken the genuine new energy revolution. There are many misleading quotes in the media, e.g., "[shale gas] is a real new revolution," "America will achieve energy independence by massively exporting natural gas," "the shale gas revolution reshaped the international energy landscape," etc.

According to statistics, shale gas production in America accounted for less than 1% of the natural gas supplied in 2000, but today the proportion has jumped to 30% and still keeps growing. Thanks to the mass development of shale gas, the US has replaced Russia as the largest producer of natural gas, accounting for 20% of natural gas production in the world. And in 2012, 71.6 billion m^3 of natural gas was sold in America, 30% more than was sold in 2006.

Many countries have started to invest more in shale gas like America has. Mexico plans to spend USD 2 billion in developing shale gas in the next two years. More than forty multinational oil companies are looking for shale gas reserves in Europe. Russia, another energy supply giant, has been preparing to exploit shale gas despite its incredibly large reserves of conventional natural gas.

Being no exception, China was shortly involved in this shale gas upsurge. In January 2012, China Petrochemical Corporation (Sinopec Group), a Chinese energy giant, bought the equity of shale gas projects from Devon Energy. After that, Sinopec and China National Petroleum Corporation (PetroChina) respectively acquired 5 shale gas fields in America and the Canada-based shale gas projects of Royal Dutch Shell for USD 2.2 billion and USD 1 billion.

Concerning the NPC&CPPCC, in March 2012, "accelerating the exploration and development of shale gas" was included in the *Report on the Work of the*

Government. Four months later, shale gas was included in the *Development Plan for National Strategic Emerging Industries during the 12th Five-Year Plan Period* and the *List of the Seven Strategic Emerging Industries* together with solar power, wind power, nuclear energy and biomass energy.

In June 2012, the Ministry of Land and Resources of PRC invited bids for the rights to exploit shale gas in four regions, which launched the shale gas "revolution" in China. In November 2012, the Ministry of Finance announced a subsidy policy: 0.4 yuan/m³ to shale gas by 2015. Only 28 months had passed since the first shale gas well in China made yields. In contrast, China had produced its first monocrystal silicon as early as 1958, but the relevant subsidy only arrived 50 years later.

The *Shale Gas Development Plan (2011–2015)* stipulates that during the 12th five year plan period, China will prove 600 billion m³ of geological reserves of shale gas and 200 billion m³ of recoverable reserves, and shale gas production will reach 6.5 billion m³ by 2015. So China is striving to increase its production to 60–100 billion m³ by 2020. If the target is attained, the self-sufficiency rate of natural gas for China will rise to 60%–70%, and the percentage of natural gas in the primary energy consumption mix will be increased to around 8%.

Table 3–1 Shale Gas Development Plan of China (2011–2015)

Proved geological reserves of shale gas	600 billion m³
Recoverable reserves	200 billion m³
Production in 2015	6.5 billion m³
Production in 2020	60–100 billion m³
Self-sufficiency rate of natural gas	60%–70%

Data source: Shale Gas Development Plan (2011–2015) *published by NEA.*

The goal is set high, but before we learn from and emulate the American experience, we should answer an important question first: Shale gas is an unconventional natural gas, more difficult to explore and with higher costs than natural gas, while its price is the same as that of other kinds of natural gas, so why then do we go after shale gas and leave natural gas untapped?

In other words, why not develop conventional natural gas first, which is easier to exploit and with lower costs, and only then move on to the more difficult and expensive shale gas?

Besides, we cannot take the American "shale gas revolution" as an absolute solution to the energy crisis, because the revolution is not completely market-based but manipulated by financial capital.

Behind the Frenzy

America's shale gas story began in 2005. On March 11, 2005, CBOT unilaterally raised the trading deposit for short sellers of natural gas contracts, aiming to encourage more transactions in the market. As a matter of fact, starting at the beginning of 2005, CBOT systematically adjusted the trading deposit for natural gas futures and relevant financial derivatives, and specifically reset the trading leverages of different short-term and long-term contracts by month, serving as reminders for transaction directions.

In layman's terms, CBOT just wanted to covertly tell everyone: come, invest in natural gas since its price is going to soar. Therefore, huge amounts of money were attracted to natural gas investment, and the unprecedented "shale gas revolution" began. An army of drillers and explorers rushed to the shale gas basins in the 48 basin states, occupying major shale gas plays in the middle and eastern parts of America, including Barnett, Fayetteville, Haynesville-Bossier, Marcellus, and Woodford.

In the same year, the price of natural gas in America was manipulated by the financial system to USD 259/kilostere, while the mass production cost of shale gas was USD 150/kilostere. That is to say, as long as it had natural gas rising from underground formations to ground level, a company could earn USD 109/kilostere. And the annual production of each well was around 10 million kilosteres, generating profits of USD 1.09 million. Therefore if you owned 10 wells, your profit would reach more than USD 10 million.

Of course, you have to invest first in order to see profits later. The cost of a well is around USD 3.5 million and 10 wells cost USD 35 million. What if you do not have enough money? Never mind, someone can lend you it in the market shaped by financial capital, so that there is always a way for you to continue playing the game. In the single year of 2006, more than 40,000 new wells were dug in mainland America. The natural gas production of America once dropped from the 526.4 billion m^3 in 2004 to 511.1 billion m^3 in 2005. However, it rebounded to 524 billion m^3 in 2006 and soared to more than 600 billion m^3 in 2010.

Mass production affected supply and demand, which resulted in falling prices. The wellhead price of natural gas was lowered to USD 225/kilostere in 2006, and fell to USD 220/kilostere in 2007. What is odd is that the wellhead price in 2008 hit the highest record for dozens of years—USD 281/kilostere, with more funds being drawn to shale gas. It was just like the oil era of Rockefeller's day.

Unfortunately, it was to be the last such spree. Market rules tell us that tragedies and glories always walk hand in hand.

Massive and disorderly development led to overcapacity. The price of shale gas moved like a roller coaster, plummeting from the lofty heights of USD 281/kilostere in 2008 to the somewhat more modest $129/kilostere in 2009, which was in fact USD 21 lower than cost. Thus, the number of wells stayed at the 50,000 in 2009 with no increase.

In 2012, when the "shale gas revolution" in China started, the one in America reached the end of its glory days, with the price of natural gas collapsing to USD 79/kilostere. Five years ago, a company could have earned USD 109/kilostere. However, today, with the same well, the company loses USD 61 for exploiting a kilostere of shale gas, and loses USD 610,000 a year. If a company owns 10 wells, it would have a net annual loss of more than USD 6 million.

Unfortunately, once launched, there is no end to the shale gas revolution empowered by Wall Street, unless a major economic crisis strikes. At the initial stage of shale gas development, most players were small but risk-taking companies who only had drills and techniques, but no funds. The investors tailored a financing plan for those small companies—buying options and selling futures, i.e., the investor first gave some cash to the company and promised to share some drilling expense, and the investors expected to get part of the future yields of the company as their payback. However, there was one prerequisite: even though the price of natural gas dropped, the drilling company had to ensure full-load operation of all the wells. Otherwise, the company would lose its equity. The reason why investors did this was to maintain a high stock price. When the price of natural gas fell to below cost, drilling companies had to keep drilling in deficit, or else they would have nothing, because all the funds had been paid to the land-owners.

According to the in-depth research and coverage of the *New York Times*, the funds from two-thirds of the American drilling companies have been locked up. Even the deep-pocketed Exxon Mobil (XOM) failed to dodge a bullet. Urged on by some investment banks, XOM acquired XTO Energy Inc., a US natural gas giant in 2010, when the gas price then was twice that of today. Rex W. Tillerson, CEO of the company, was so furious that he ordered the owners of wells to stop drilling and shut down wells. At an industrial summit in New York, he admitted in public: "We are all losing our shirts today. You know, we're making no money. It's all in the red."

As a matter of fact, many people were critical of shale gas. Jean Laherrere, co-founder of the Association for the Study of Peak Oil and Gas (ASPOG, France), once said that shale gas was more of a "Ponzi scheme" than a revolution.[2] Oil companies exaggerated the size of natural gas reserves in order to make their balance sheets look good.

What's more, an anti-fossil-fuel think-tank called the Post-Carbon Institute (PCI) published a report pointing out that shale gas was nothing but an interim solution. It even uploaded the report outline to an interesting website—shalebubble.org—whose name says it all.

The press also keeps questioning shale gas. In 2012, the *New York Times* quoted several internal emails from insiders in the industry and the geological sector, questioning the motives of oil companies who overestimated the prospects for shale gas and overstated the reserves.

The *Hong Kong Economic Times* had a similar report, which said that oil companies unanimously boasted about shale gas but shunned the topic of its difficult exploitation and high processing costs. There are lots of active gas wells in America, but they are usually surrounded by many more wells that are unproductive and costly to develop. Besides, the production of major wells falls faster than people think. Contractors believe that shale gas is "unprofitable in nature," while independent commenters with no vested interests in the shale gas industry take it to be a huge "Ponzi scheme."

There are even doubts from within oil companies themselves. Having huge doubts over the unexpectedly fast decline of well production following the first years, a geologist from the Chesapeake Energy Corporation, the second largest natural gas producer in America, pointed out that the production forecast of the company was unreliable. An employee of ConocoPhillips Company predicted that shale gas would end up as the least lucrative business in the world.

However, some oil companies are still obsessed with shale gas, and continue to lobby investors to believe that shale gas is the oil of the future. But many insiders remain unconvinced, because exploiting shale gas is still merely an emerging industry. The reserves are only speculation based on limited data. Some even question whether oil companies exaggerate the reserves on purpose in order to "cheat Wall Street" by whitewashing their performance with the most productive wells.

So then shall we put a question mark on the shale gas boom in China? Putting aside the long-term value of shale as a resource, we should let market rules determine our energy future and the route to energy independence.

Shale gas development in China is still in its infancy, so it still has time to reconsider or adjust its investment strategies. Because shale gas reserves are another bank of natural gas, and China does need related technologies, it should develop them with discretion, rather than simply following the Americans.

Lacking experience and still only possessing immature technologies are restraints to the development of shale gas in China. There is still a long way to go before the mass-development of this emerging and unconventional resource.

The Truth About Shale Gas

Shale gas is an unconventional natural gas, but it does still fall into the category of fossil energies. The discovery and development of shale gas may secure the throne among fossil energies, and postpone the day when they are replaced on a large scale. However, it makes no sense to equate the "new energy revolution" with the "shale gas revolution," which does not fit the criteria of "being renewable and pollution-free" set by the UNEP. New energies must both address energy demands and enable sustainability.

Currently, the core technologies of shale gas exploitation are horizontal drilling and hydraulic fracturing. Half of the shale gas reserves in America are mined with the latter approach, undermining people's health and the environment, with air, water and soil pollution. And there is another inherent contradiction: regions where shale gas is produced are usually short on water, and the industry consumes lots of water.

Because drills go through an aquifer, the chemical additives used in drilling will pollute the nearby groundwater. Taking America as an example, in May 2010, the Keystone pipeline, carrying heavy oil, had two major leaks, which caused severe pollution and ecological damage. It was due to the wide variety of dangers that US environmental, agricultural and husbandry groups opposed developing unconventional oil and gas. For the same reason, the Obama administration postponed the extension of the Keystone pipeline.

Research carried out in Pennsylvania has also proved that shale gas is a threat to drinking water. It showed that drinking water within a one-kilometer radius of a well may be polluted by methane, ethane or propane emitted during exploitation.

Concerned about geological disasters, the governments of the states of New York and Delaware have halted shale gas projects in their regions. The media even warned that although exploitation gave rise to "cheap fossil fuels" in America again, short or mid-term profits might drag the country towards a long-term dependence on fossil energies.

In addition, whether this US "revolution" is replicable in other regions is still a question. According to a report released by the Urban Financial Research Institute of the Industrial and Commercial Bank of China (ICBC), there are many obstacles in the way of shale gas development worldwide. First, different geological conditions in each region will make development harder. Second, in exploiting shale gas, much water will be consumed, and chemicals involved in the process may pollute local groundwater. Third, an extended period of low prices discourages companies from investing. And finally, in the markets beyond North America, there is insufficient infrastructure to store, liquefy and transfer unconventional natural gas, which constrains development in those markets.

All the relevant data suggests that shale gas reserves in China are as big as those in America, and China is a large energy consumer, second only to the US. But if we stay calm, we can see that China cannot put up with the environmental consequences caused by shale gas development given China's existing resources.

The land and water resources per capita in China are much lower than those of the US, and its environment is more vulnerable. On average, a well of shale gas requires 200,000 tons of water. The fracturing fluid injected into the shale is full of chemicals, bringing about uncertain effects on groundwater. And shale plays in China are mainly inland areas and basins where water is scarce.[3]

In the "12th five-year plan period," when we exploit shale gas in water-scarce regions with immense resource potential like North China, the Zhungeer Basin, the Tuha Basin, and the Eerduosi Basin, we can't be too careful. Industrial experts remind us that in the mass development of shale gas in Liaoning, Shannxi and Sichuan, industry and agriculture will probably compete for water.

More importantly, China has no qualified technology for exploiting shale gas. Nowadays, mature technologies are based on the drilling techniques accumulated from specific regions and geological conditions in America. It is impossible for us to bring in the American technologies quickly. Even if we succeed in doing so, we are not sure whether they are suitable for the unique geology and environment in China, let alone what their huge exploitation cost might be.

Mr. Zha Quanheng, an expert and former Deputy Director of the Exploration Department under the Ministry of Oil, wrote an article saying that if China could not find a "more land-, capital-, water-, environment-saving" way to develop it than the American model, then China should not promote the "shale gas revolution" quickly, in case of following the US blindly.

Putting accusations of it being a "Ponzi scheme" aside, we cannot say that shale gas represents the vanguard, since it is a fossil energy. In the long run, the authentic energy "revolution" should coincide with epoch-making technological innovation. That's why we believe that the next energy transformation will happen in the realm of solar power, after technologies are continuously improved and costs rapidly fall, instead of in shale gas with its uncertain prospects.

Chinese energy companies must keep a sober mind in the face of various "revolution" theories. And the biggest challenge is how to define the big picture for energy development and be prepared in advance, rather than to follow some "complicated trend."

PV Goes Astray in the US

There is a saying "Throw a stone anywhere around Capitol Hill, and you'll definitely hit a couple of lobbyists."

It means that, like many others, consortiums on behalf of traditional energies send lobbyists to Washington. In their own interest they even assign their staff to work in government in order to manipulate politics in a "revolving door" manner.

Because of this, the new energy policies of America have been swinging back and forth since the Nixon administration 40 years ago up to the current Obama administration. As a result, the US has been outpaced by others in solar power development. It has no advantage compared with China.

Yet we should never belittle America's strengths in solar power. It lacks neither investors who are interested in solar power, nor drive towards technological innovation.

Political Twists and Turns

It is only several hundred meters from the Willard InterContinental Washington to Capitol Hill, where congressmen are often stopped on the way to the office and told by lobbyists that traditional energies need more funding and policy support from the federal government.

Such lobbyists are employed by a variety of sectors, with many of them serving fossil energy companies in the US, and they naturally often lobby against new energies. *The Economist* has repeated the expression "Throw a stone anywhere around Capitol Hill, and you'll definitely hit a couple of lobbyists."

Consortiums of traditional energies manipulate government policies with their economic clout. So it is understandable for the US government to have wavering PV policies. In the past, the US government paid a lot of attention to the solar power revolution, and promoted a "Million Rooftop Program." Moreover, President Obama was a big fan of solar power in his first term, talking about First Solar, Inc. (FSLR) more than a little. However, America is a country manipulated by mega-corporations influenced by intricate benefit networks. Shale gas let the US see some hope. As a result, the US takes a conservative approach towards mass solar development. In this way, solar development in America underwent a series of twists and turns.

The first twist-and-turn occurred 40 years ago when then-President Richard Nixon aspired to energy diversity. And when President Carter took office, he proposed opening a solar power bank in order to reduce excessive dependence on oil. He also promised to "make 20% of the electricity from solar power by 2000," and requested the installation of solar panels on the rooftops of the White House. However, his successor, President Reagan, had little interest in solar power and gave orders to remove all the solar panels Carter had built soon after he took office. In this way, solar power development was suspended.

The second twist-and-turn occurred in early 1990s. With the beginning of the Gulf War, the world faced shortages in oil supply. Therefore, President George H. W. Bush put the solar power program of President Carter back on the agenda and upgraded the Solar Power Institute to the National Renewable Energy Laboratory (NREL). However, after Saddam retreated from Kuwait, falling oil prices left the program stranded once again.

The third twist-and-turn occurred in and after 1997. On June 26, 1997, President Clinton submitted a report to Congress on environmental development, putting forward the "Million Rooftop Program." It was to be a mid and long-term initiative in the 21st century, promoted and implemented by the US government, and was also the largest solar power project at the time. The government expected to install solar power systems on 1 million rooftops by 2010. The functions of the system included PV power generation, and water and air heat collection. Unfortunately, the program did not work well and so solar power was not massively developed or utilized. In 2000, the installed capacity of solar power in America was exceeded by that of Germany. And at the beginning of 21st century, when George W. Bush rose to power, he declared the US's withdrawal from the *Kyoto Protocol*. Consequently, America was outrun by most all European countries in solar power utilization.

The development of new energies in America has not gone smoothly since then. In 2005, the US Congress passed the *Energy Policy Act (EPAct) of 2005*, which has been continually criticized. For one thing, there are flaws in the regulations on energy conservation, efficiency improvement and the promotion of renewables. For another, EPAct didn't offer solutions to the core challenges of US energy policy, chiefly how to address climate change and energy security. Besides, the act gave priority to fossil energies and nuclear development again, which meant that the US energy policy didn't shift its "focus on fossil energies."

When President Obama took office in 2009, he advocated a new energy revolution, and Americans emphasized the PV industry again. In the first year of his term, the installed capacity of solar power grew to 0.43GW from 0.36GW in 2008, and then jumped to 0.87GW in 2010, 1.85GW in 2011, and 3.6GW in 2012. However, no one at that time foresaw the good momentum of solar power being sabotaged by shale gas in mid stream.

In recent years, to reduce oil imports and dependence on foreign energies, and to achieve "energy independence," America has launched a dramatic "shale gas revolution" which is more sizzling than the solar boom.

Superficially, America has done a good job in developing solar power in recent years according to the statistics. Around 2011, the installed capacity of solar power grew fast partially because the preferential taxes for this sector were about to expire. That is to say, the prosperity in 2011 was only an overdraft taken from future growth.

Table 3–2 Top 5 countries with Cumulative installed PV capacity in 2012

Country	Cumulative installed capacity in 2012 (MW)
Germany	32 420
Italy	16 250
China	8 250
USA	7 583
Japan	6 551

Data source: Global Renewable Energy Report (2013)

For the American PV industry, the end of 2011 was a turning point. Back then, supporting approaches from Obama's green policies, e.g., subsidies and debt guarantees, were due in succession. The only one that still existed in 2012 was the Production Tax Credit (PTC). Against this backdrop, with more and more investors transferring their money to foreign markets and companies downsizing employees, the idea that "Obama's green policies are ineffective" came to prevail in the world's mindset.

The US government seems to attach great importance to developing new energies. However, its fluctuating polices suggest that there is no unanimous, continuous or systematic strategy for new energies there. The root cause of this is that America does not believe that new energies will bring major breakthroughs in the foreseeable future.

The US Department of Energy (DOE) believed that only by lowering the building cost of a solar power station to USD 1/watt, could the stations remain competitive after the subsidies were lifted. However, by the end of 2011, the cost was still as high as USD 4.08/watt, much higher than the threshold DOE cited.

Unconventional energies like shale gas have developed rapidly in recent years. However, at such small scale and as immature technology, it has met barriers to sustaining that development. Besides, with the mass development of shale gas, especially its huge growth in production, prices fell dramatically. As a result, the new energy industry that mainly relied on government subsidies was less competitive.

The "Revolving Door" Effect

In spite of the twists and turns, many American investors are still not only interested in the solar market, but also keeping an eye on the market and seeking evidence that energy substitution can be lucrative. Nonetheless, they also know that their failure or success depends on politicians and politics.

Unfortunately, the current rules of American politics are unfavorable to new energy entrepreneurs, whose opponents are traditional energy oligarchs. With slim chances of winning and uncertain policies, it seems inevitable for them to fail, even with the best ideas.

They envy how the semi-conductor and Internet sectors develop, where revolutionaries can play fairly and equally, while they in the energy sector have to fight against the interests of mighty traditional energy tycoons.

Traditional energy companies are powerful, not only because of their big size, but also their huge influence on making national energy strategy through lobbying. Statistics show that oil and natural gas companies spend around USD 60 million every year on lobbying, an activity aimed at influencing law and decision making in order to gain specific benefits. The Government Accountability Office of the US Congress revealed that these big companies can earn USD 6 billion each year by lobbying. To solar companies, USD 60 million can represent an impossible amount of capital, while to fossil energy companies it is only a drop in the bucket, less than the daily revenue of EXOM. More importantly, the money spent on lobbying can mean enormous returns for the interested parties.[4]

Besides the impact of lobbying, the wavering solar power policies of America is also connected with bipartisanship and the presidential system. The Republicans and Democrats rule the country in turn after close races, the president takes responsibility for his or her party, and the Cabinet takes responsibility for the president. But small parties cannot realistically compete in the presidential election. Consequently, when a party wins the election, the interest groups it represented get the power to make national policies. A newly-elected president can abolish or modify the policies made by his or her predecessor, and so there is little or no policy consistency.

The Democrats represent the core interests of the middle-class in America, which cares about better environmental and living standards, and this is why

Presidents Carter, Clinton and Obama are supporters of solar power. However, the application of solar power inevitably hurts the interests of large companies. Therefore, it's normal that Republican Presidents Reagan, G. H. Bush, and G. W. Bush, who represented large companies, showed no interest in developing solar power. As president, they received huge donations from the interest groups they represented when they ran for president.

To reinforce lobbying, large companies usually recommend their staff for positions in government departments, and such switching of role from corporate executive to government official is called the "revolving door" effect.

Even the pro-solar Obama failed to keep away from lobbyists. When the installed capacity of solar power kept growing in the first term of the Obama administration, fossil energy companies took countermeasures—for instance, America focused on developing large-scale concentrated power stations, where the power would be transmitted by traditional energy groups from sparsely populated regions to the densely populated east coast of America via ultra-high voltage grids. However, such concentrated production and distribution of renewables was not recognized by the governors and power companies in the East. In July 2010, governors of eleven states in New England and the Mid Atlantic region wrote a letter to Harry Reid (leader of the majority party) and Mitch McConnell (leader of the minority party) of the Senate, publicly opposing the national power transmission policy. The governors maintained that concentrated power generation from wind and solar in the west "would jeopardize the local efforts of generating power with renewable energies and do harm to creating employment related to clean energies in the state." Fourteen power companies (many of which suffered business losses in those regions) together with the governors asked Congress to allow those regions to develop renewables independently.[5]

In addition, US solar policies are correlated to changing oil prices. Solar power generation will boom when oil prices and subsidies are high, and collapse when oil prices fall and subsidies get canceled, as they did during Obama's second term.

The development of solar power in the US has proved the significance of consistency in government policies. With fair policies, it is not government but the impartial market without any hidden subsidies that would choose the winners. A fair game (badly needed by the risk-taking investors and technological revolutionaries) requires no guarantee of profits, but fully realized competition. Even given this, many players still failed, and they claimed, "The stakes and costs are unbelievably high."

America Still Matters

As commentators have said, if America wants to develop its green energies, it has to leverage the impulse of capitalism for profits and let the market play its role.

Although the investment tax credit (ITC) empowered solar development, insiders took it to be more of a stopgap solution, doing no good for long-term growth. They believe that only with steady tax policies can investors work out their long-term investment strategies, and mobilizing the capital market is the best way to promote solar technologies rapidly.

Therefore, it's impossible to develop solar power on a large scale in the US only through the efforts of a bunch of environmentalists. Instead, a large amount of funds is required to phase out coal and oil. Quoting Vijay Vaitheeswaran, a journalist for *The Economist*, "the practical approach is marrying technology to money."

Rich in sunshine, California ranks top among all States in playing host to ideal conditions for solar development. It passed a law to determine and limit the total amount of carbon emissions, bringing about big changes in that area. With that, California disproved the doubts of some organizations on solar power and even likely helped the whole industry attract more funds. Some may call it a bubble, but I think it is a boom. On January 10, 2009, California decided to invest USD 3.2 billion in the "Million Solar Rooftop Program," planning to build a 3GW solar power generation system, and reduce GHG emissions by 3 million tons.

Legislation in California has proven that proper laws generate numerous job opportunities, improve social welfare, and greatly promote international competitiveness. It has also refuted the claim that carbon laws would destroy the economy.

The American interpretation of a fair and sound competitive market is that with carbon tax and tax alike, the real prices of fossil energies will be reflected and "the market will make its choice."

In the meantime, the legislature required the cancellation of huge subsidies to traditional energies, to ensure market fairness. Among the top 10 companies in the world, seven are oil companies, but fossil fuel producers, the seven included, still receive taxpayer subsidies in America. In 2010, financial support from the US government to these companies registered USD 15.4 billion. According to a report from DBL Investors, subsidies from the federal government for oil and natural gas were five times that given to renewables in 15 years.

If market fairness is achieved, making renewable energy investments actually profitable, PV development in America will be successful. With existing industrial and policy strengths, the American PV industry will be very competitive.

As the cradle of PV technology and industry, the US has always led the advance and commercialization of PV technologies. In recent years, the US has worked out a national PV development plan, which will greatly boost technical progress in this sector. By partnering with universities, laboratories, and industrial venture investment organizations, the plan is aimed at making short-, middle-, and long-term strategy plans and targets for technological advancement and application via collaboration. The long-term goal is to reduce the cost of PV power generation in America to 6 cents/kWh.

Technological advancement and new energy plans have empowered the rapid growth of installed PV capacity in America since 2012. Installed PV capacity increased by 3.6GW, with a year-on-year growth rate of 92%, and the cumulative installed capacity reached 7.6GW. The *Renewable Portfolio Standards* (RPS) of different states and the ITC guaranteed the continuous growth of the PV market. It is estimated that the installed PV capacity of America would increase to more than 4GW in 2013.

We care about the influence American PV technologies exert on competition because we know that the one who holds core technology leads the pack. Of course, the fast recovery of the US market does no harm to the Chinese PV industry. Many PV professionals believe that Obama beginning his second term means that the US application market will be further opened, which is good news for Chinese PV companies. The anti-subsidy tax and anti-dumping tax collected by America only targets crystalline silicon cells made by the Chinese. If Chinese companies switch to producing thin-film panels, they are likely to return to and dominate the US market.

It is time for China to take thin-film solar technology seriously. At present, major developed countries (America, Germany, and Japan) are all producing PV cell components with thin-film technologies, and most of China's core equipment for thin-film production is made in those countries. No one else can produce crystalline silicon cells at low cost and on as large a scale as China does. But thin-film solar products are less likely to face anti-dumping and anti-subsidies investigations in the current competitive sphere, because the mainstream thin-film technologies vary in conversion rate and cost, but for the same technology, there is little price difference among countries with different labor costs, thus there is no room for dumping at lower prices. Therefore there will be no anti-dumping and anti-subsidies problems.

At the same time, the advanced equipment required for thin-film production in China is not made domestically, yet imported from abroad. If the manufacturing countries impose "anti-dumping" investigations on Chinese thin-film products, they will actually be opposing and hurting themselves. China can become a leading producer in thin-film cells for its technology and production scale. It can still develop products with high technology, excellent quality, and for reasonable profit and prices.

What's Wrong with German PV

How did the "Sunflower House," the German version of the "1,000,000 Solar Rooftop Program" come into being?

Germany passed the Electricity Feed-In Act in 1991, the Regulations on Electricity Grid Integration of Renewable Energy, the Ecological Tax Reform Law in 1999, and the Renewable Energy Law (EEG) in 2000. In 2010, Merkel shut down 17 nuclear stations. Germany did a good job in developing PV because it is protected by law, and supported by strong politicians.

"All solar companies with high market value make profits by selling their products to Germany." Then why was Germany first to initiate anti-dumping and anti-subsidies investigations against Chinese PV products? On the surface, Germany seemed to cut its solar subsidies due to the European debt crisis. Actually, it was because the German PV industry neglected the pursuit of innovation in technology, and thus slowed down development.

How were the "Sunflower Houses" Built?

You will see many houses with solar panels in Germany, and these are called "Sunflower Houses." As the leader of the BIPV, Germany is like a magnet for participants in the PV industry.

Most of the people of Vauban in Freiburg, in the southwest of Germany, are living in "Sunflower Houses." These houses are 4 stories high, and have a cylindrical shape. The buildings have solar panels on their rooftops, which slowly move to automatically follow the Sun from the best angle. They can generate sufficient energy even in winters with weak sunshine. The energy, besides being utilized by the residence, is also sold back to the power grid, with an annual profit of around €6,000.

Thanks to subsidies, a "Sunflower House" can supply power to an individual home and create profit. Germany is an active advocate for solar power, and it also applies solar technologies of the largest scale in the world. However, the development trend has somewhat changed, recently.

Since 2012, Germany has slowed down its PV development. For one thing, the government slashed PV subsidies. For another, many local PV companies went bankrupt. Is it a policy shift or the European debt crisis that distracted Germany from the new energy revolution?

At the end of 1990, Germany announced its "1,000 Rooftop Program," to install 1,000 PV power generation systems of 1–5 kW on residential buildings within 3 years. This would validate whether integrating a solar power system into power grids was economically and technically feasible, and so whether it is in fact an available alternative power source. The target was later raised to 2,500 sets, and from then on the German PV industry has been quickly developing.

In 1991, Germany enforced the *Electricity Feed-In Act*, which defines the on-grid prices of wind and hydro energy as 90% of the retail tariff. Therefore, public utilities had to buy wind and hydro electricity at these prices. The electricity generated must be integrated into the grids, which should in turn buy all the surplus power at the defined prices. In the same year, the government issued the *Regulations on the Integration of Renewable Energy Electricity into Grids*. It stipulated that renewable energy power generation had to be integrated into grids, a lowest feed-in rate had to be established, and that a subsidy of 0.199Dm/kWh would be given over 20 years.

In January 1999, Germany began to implement the "100,000 Solar Rooftop Program" on a larger scale. By 2003, 100,000 rooftop PV systems would be installed, with a total capacity of 3–5 million kW. To ensure the smooth execution of the project, the Federal Ministry of Economics and Technology earmarked a fiscal budget of €460 million. Germany also promulgated the first *Renewable Energy Act* in 2000, which was amended in August 2004, January 2009 and 2011. With feed-in tariffs, the German government provided a 20-year subsidy of from €0.45 to €0.62/kWh, based on different types of solar power generation, with the subsidy reduced by 5%–6.5% each year.

The feed-in tariff catalyzed the growth of the German PV market. Within only a few years of the act being issued, the German PV market began to develop rapidly. Many public and private buildings were installed with solar panels. In 2004, Germany added 363,000kW of solar power generation systems, an increase 2.35 times over that of 2003. With a growth rate faster than the average speed (59% per year) of the global PV market, Germany replaced Japan as the largest PV market. From 2000 to 2005, a bunch of PV companies, like Q-Cells and Solar World, emerged as world-leaders thanks to an annual growth of 38% for the whole PV market.

In addition to legal and policy support, the German government has provided financial incentives for the renewable energy industry. For example, KfW (formerly KfW Bankengruppe), a German government-owned development bank, extended loans to PV producers and renewable energy companies with a 50% discount in interest rate. In 2008, KfW provided €340 million in funds to renewable energy-related bodies globally (not including large hydropower stations), making itself the largest financing institution for renewable energies.

What's more, the Deutsche Energie-Agentur GmbH (Dena) —the German Energy Agency—also offers consulting services and financial support. Dena is not a government agency, but a company jointly controlled by the German government, KfW, Allianz Group, Deutsche Bank, and Deutsche Bundesbank. On its websites, one can find governmental information on renewable energies from both the EU and Germany.

Local government throughout the country either issued regulations and policies for their own states, or launched promotion activities for renewable energies in certain sectors, in order to boost the regional development of new energies.

A series of government support programs enabled the soaring growth of Germany's installed PV capacity. According to the Federal Statistical Office of Germany, the share of renewable energies in total power generation rose from 16.4% in 2010 to 22.1% in 2012, while the share of nuclear power dropped from 22.4% to 16.1%. That is to say, renewable energies have already outranked nuclear energy as the second largest power source in that country.

Table 3–3 Share of power from renewable energies in total power usage in Germany

Year	Share(%)
1990	3.1
1995	4.5
2000	6.4
2001	6.7
2002	7.8
2003	7.5
2004	9.2
2005	10.1
2006	11.6
2007	14.3
2008	15.1
2009	16.4
2010	17.1
2011	20.0

Data source: Zhang Qin, Zhou Dequn: Research on The Development of New Energy Industry in China, Science Press, published in 2013.

Incumbent Chancellor Angela Merkel is a key figure for the growth of renewable energies in Germany. In August 2010, her administration declared the closing of 17 nuclear plants over the next 10–15 years, and required the unclosed ones to pay €2.3 billion in taxes every year to support renewables—solar power in particular.

The decision of her administration offended some in economic circles. Many celebrities and groups including Deutsch Bank and Metro Co., Ltd. wrote letters to oppose Merkel's energy policies publicly. They called for resuming the usage of coal and nuclear energy, claiming that abandoning these two resources too early could lead to a loss worth billions of euros, and they also attacked the tax on nuclear fuel. Four grid operators warned Merkel that her decision to shut down nuclear reactors would cause national blackouts in large areas in winter, and in days without sufficient sunlight or wind. Relying only on wind or solar power generation would lead to power failures in the regions in southern Germany with heavy industry, they claimed.

Hans-Peter Keitel, President of the Federation of German Industry (a major industrial lobbying organization) said: "If (power supply) can no longer be guaranteed, it will weaken Germany as an industrial nation. Politicians must guarantee the stability of the grid and the system during this transformation in energy policy."

In the face of these threats and warnings, Merkel did not flinch. On the contrary, she fought back even harder by saying that the threats and warnings only served to make the German government firmer in its support of the new energy industry.

Why Germany Slowed Down

With Merkel's support, German solar power generation kept developing rapidly for many years. However, external uncertainties cannot be controlled by her will alone, and the momentum is no longer there.

In 2012, Germany installed another 7.6GW of PV capacity, and the total cumulative installed capacity reached 32.4GW, ranking top of the world. The data on the installed capacity in 2010 and 2011 shows that PV growth in Germany slowed down. The newly installed PV capacity of Germany was 7.406GW in 2010 and 7.5GW in 2011. The increase in 2012 was 100MW compared with that in 2011.

An amended policy on feed-in tariffs was believed to be the main reason for this slowdown in 2012. As a matter of fact, because the PV power generation market was becoming increasingly mature, the installation cost gradually diminished and so when the economic crisis broke out, the German government reduced on-grid prices and subsidies at a faster rate year on year from 2008 onwards. Instead, it adopted a policy of supporting scaled development instead of subsidizing the cost.

The *2012 Amendement of the Renewable Energy Act* further stipulated that the on-grid PV prices would be cut based on the newly installed PV capacity of the year. The benchmark cut rate was set at 9%. If the installed capacity of a year exceeds the annual quota of 3.5 million kW, the on-grid price will be reduced by another 3% for each 1 million kW beyond the quota, with a maximum cut of 24%.

> Visionary and reasonable policies plus strong support from the government to develop and utilize new energies made Germany, which used to be the leader in PV research and manufacturing, the largest PV market in the world. Today, however, Germany has quite different PV policies and so has slashed subsidies, providing insufficient incentives to companies. This has shaken its leadership in this sector.

Also, the German government declared in June 2012 that it was going to cut down PV subsidies by 20%–25% every year from 2013, stop subsidizing rooftop installation from 2015, and terminate solar subsidies in 2017.

While the Germans were axing subsidies for PV companies, their Chinese peers emerged quickly to seize global markets. As a result, the whole PV industry in Germany slipped into a slump, and many large PV companies were kicked out of the top 10 list in the industry.

In 2013, a lot of German PV companies went bankrupt, with the number of jobs shrinking from 110,000 in 2011 to 87,000, and overall sales plummeting to €7.34 billion, €4.56 billion less than that of 2011.

Many people believed the European debt crisis triggered the slash in subsidies, and made the competition more intense between German PV companies (e.g., Solar World) and Chinese ones. Finally, it led to the anti-dumping and anti-subsidies investigations.

As the largest PV company in Germany, Solar World proposed to America and the EU on October 19, 2011 and July 24, 2012, respectively, that they conduct anti-dumping and anti-subsidies investigations on crystalline silicon PV products from China. In fact, it is not only the crystalline silicon cell producers who are troubled by competition, but also thin-film technology which is the technical mainstream and a common challenge to all crystalline silicon manufacturers. Even if results from the investigations are favorable, it is not likely that Solar World will remain invincible.

The German government did not blame change in demand, but the abrupt drop in price of PV systems for the falling turnover of the PV industry. This is because from 2010–2012 the annual installed capacity remained the same. However, in that period, the PV market did not shrink.

Looking at the status quo of German PV companies, their executives believe governments should transform their policies. Until now, governments have promoted renewable energies by subsidies and orders, which help, to some extent, to boost the development of renewable energies. However, these are not the most reasonable of approaches. For instance, Japan and Germany only make up 10% of the global energy market, but account for 70% of the world solar market with long-term subsidies. In 2007, German power companies delivered ¢57/kWh to owners of solar power generation systems, almost three times the electricity rate sold to consumers. That is why that policy pushed other countries to follow suit in investing hugely in solar products. "Companies in the world all sell their products to Germany to get high market value." Government subsidies and administrative orders have fatal flaws. For example, if a government is required to determine the target and amount of subsidies, it has to precisely master technologies and forecast their trends, which goes beyond the capability of government officials.

In addition to decreased subsidies, stagnant technology also undermined the competitiveness of German PV companies. In a report, the German government admitted that many PV companies went bankrupt because they "ignored technical innovation." The government encouraged PV companies to establish partnerships with suppliers and research institutions to maintain their market positions.

A commentary in the *Süddeutsche Zeitung* on August 7, 2013 noted that the German and EU PV industry in general lost to their Chinese counterparts because they moved slowly in the field of technological innovation. So the key to their survival is to maintain leading positions in technology.

The article also mentioned that formerly-leading German companies have fallen behind, endangered by their declining technological competitiveness. If they don't change this, they can hardly be expected to survive. So they have to transform themselves and focus on R&D and innovation.

If they do not change, shrewd investors will choose Asian factories that boast larger scale production, newer equipment and lower production cost over Saxony or the State of Brandenburg.

Indeed, compared with Asian manufacturers, German PV companies are not only slow at innovation but also lacking large-scale production capacity. Technological upgrades usually begin with high costs, but these costs will significantly drop until the upgrading is down to a certain scale.

The evolution of solar power tells us that if we want to expand the PV market, we must lower the cost with scale economy and eventually get demand to grow; we have to invest massively and promote productivity quickly in order to reduce production costs and prices sooner rather than later. However, the market German PV vendors occupy is not big enough for large PV companies to grow in, which means Germany is at a disadvantage in large-scale production.

And this is exactly the root cause for changes in the German PV market in recent years.

The Achilles Heel of the EU

The decline of the German PV industry has already disturbed the EU market. The funding for the PV industry in the EU dropped for the first time in 2012 after 8 consecutive years of growth. It fell from USD 70.14 billion in 2011 to USD 51.36 billion, more than USD 10 billion lower than the USD 63.34 billion in 2010.

And the EU investigations into Chinese PV products only made the EU counterparts more worried about the future. On May 24, 2013, the initial stage of the EU's sanction against China, Alliance for Affordable Solar Energy (AFASE), an industry association of more than 580 European PV companies, held a "funeral for the European PV industry" in the headquarters of the EU in Brussels. It aimed at attacking the European Commission for not considering the interests of the whole PV sector in the EU in adopting the investigations. Despite a reconciliation between the EU and China, the decline in the PV manufacturing industry in the EU has not been reversed.

EU PV companies started to go bankrupt in 2012. On September 6, 2012, Soliker, a Spanish BIPV company, filed for bankruptcy. On October 10, Yohkon Energía, a Spanish PV component manufacturer, filed voluntary bankruptcy proceedings at the Commercial Court number 1 of Valladolid. In the same month, Ecoware SpA, an Italian PV company, began bankruptcy proceedings. And two German solar companies, Conergy and Gehrlicher Solar, also filed for bankruptcy in 2013, during June and July respectively.

> The main challenges faced by the EU PV industry include high costs for labor and production factors, low efficiency, slow R&D, and a funding shortage. Besides, EU PV companies lack entrepreneurship which dampens their competitiveness.

Along with the adjustment in subsidy policies, several factors cause these companies to face losses.

First of all, high costs for labor and production factors: this is a common problem for the European manufacturing industry. With the high prices of PV components and with subsidies, European PV companies still hold on to some of the market. However, with escalated competition and reduced subsidies, high production costs become a constraint for these companies. According to research on 300 solar panel manufacturers, 88 of them plan to shut down their plants in Europe and North America within two years, because the production costs in these regions are too high for them to win against the competition.

Second, low efficiency and slow R&D. Low efficiency is commonly seen in all of Europe and has even become something of a "culture" there. When they are asked to arrange time for work and for their personal life, most European people choose to have more leisure time. That has slowed down economic growth and further widened its productivity gap with America, and even some Asian countries. It is also true in the EU PV sector. Being inefficient and having no sense of urgency made European countries less competitive in the field of technology compared to other countries. Since technology serves as the pillar of PV manufacturing, operations and the market, and so is a core factor in achieving competitiveness, the EU has fallen far behind China.

Third, the funding shortage. Haunted by the debt crisis, most European banks prefer not to give new loans to the "capital-intensive" PV companies and expect them to pay back previous loans. Therefore, PV companies with financial problems have no access to external support, and even internal financial restructuring can be a dead-end approach. Taking Gehrlicher Solar, a German company that filed for bankruptcy in June 2013, as an example, it went insolvent because a bank terminated a loan of USD 109.5 million.

Compared with German companies, Spanish PV makers are even more financially beleaguered. According to the Spanish press, in June 2013 Dongfang Electric Corporation (DEC) and China Export&Credit Insurance Corporation (Sinosure) required involuntary bankruptcy proceedings be started on Solaria Energia y Medio Ambiente SA (hereinafter referred to as "Solaria"), a Spanish PV component manufacturer and project developer, because Solaria owed more than €3.1 million to DEC. The Chinese party declared that it was necessary to start the proceedings, because the working capital of Solaria was negative €30.9 million.

The Association of Renewable Energy Producers and Investors (ANPIER, Spain) warned recently that, being extremely short on cash flow, Spanish PV companies have accumulated heavy debts and can find no new channel to refinance. If the government collects energy tax on this basis, 80% of Spanish PV manufacturers will go bankrupt.

Last but not least, EU companies do not have the necessary entrepreneurship, which in turn lowers their competitiveness. The theme of the 11th European Business Summit in May 2013, was "Unlocking Industrial Opportunities—An EU Strategy for Competitiveness." Representatives believed that it was lack of entrepreneurship that has seen the EU barely revitalize its industries at all recently. EU companies haven't rebuilt their confidence even as the economy recovered. Due to the absence of pioneering spirit, the recovery in Europe has been contained, and the EU has had its Waterloo in the new round of PV competition.

What's Next, Following the Awakening of South Korea and Japan?

Nuclear shadows and the scarcity of coal and petroleum forced Japan to at last choose solar power. After a temporary setback, Japan's solar PV industry witnesses a re-awakening. And with the government's strategic support, South Korea's PV industry aims to be among the world's top five in 2015.

Naturally, Japan's closed-door development of its PV industry has effectively protected its core technology from flowing out, but it also makes it hard for it to make breakthroughs in core technology. Although focusing on polycrystalline silicon, South Korea's PV industry also suffers from overcapacity, and now is struggling to reverse the situation.

As such, in the solar photovoltaic (PV) industry, we need to attach a fair amount of importance to Japan and South Korea as competitors, but it's unnecessary to be afraid of them.

Japan's Revival

The shadow of nuclear radiation has haunted the Japanese people for years. The Fukushima nuclear disaster on March 11th of 2011 severely strained their already frayed nerves. Kenzaburō Ōe, a Nobel laureate, called on the Japanese government to close nuclear power plants, by saying, "the Japanese people should be aware that the catastrophic consequences this nuclear accident had on the human body are no less serious than the American bombings during World War II."

Masayoshi Son, Japan's richest man and Softbank chairman, also joined the anti-nuclear group. After the Fukushima nuclear accident, he declared that he would donate one-third of his fortune to the disaster-stricken areas, and promote solar energy development by building ten solar power plants.[6]

For resource-poor Japan, nuclear power is almost the only viable way to achieve self-sufficiency in energy supply. Actually, Japan has extremely mixed feelings about nuclear power. To develop thermal power would doom the country to destroying their pleasant environment protected through years of effort, and severely threaten the country's energy security. However, a reliance on oil and coal means Japan's energy supply chain may break at a difficult time, and Japan's large and complex industrial system might collapse.

Thus, Japan became the first country to support the development of the PV industry by making relevant developmental policies. In 1990, Japan modified relevant technical specifications and requirements in the "Electricity Utilities Industry Law" to promote the application and popularity of grid-connected

PV generation. Two years later, the Japanese government formulated a "New Sunshine Program" whose basic objective was to support new energies as an important way of procuring energy. Within the framework of this program, the Japanese government stipulated that the government would subsidize residents who install solar PV systems by 50% (decreasing yearly) of the cost from 1994 onwards. The government would purchase solar power sent by homes at the selling price of the grids.

With these supportive policies, Japan became one of the most important PV markets in the world. Based on the successful "New Sunshine Program," the Japanese government formulated an "Advanced PV Generation Plan" (APVGP) in 2001. This plan gave priority to reduction in importing oil (accounting for 53% of Japan's energy consumption) and keeping its commitment to cutting greenhouse gas emissions in the Kyoto Protocol.

In order to promote technologies related to energy efficiency and the application of renewables such as PV and wind power, Japan introduced the *Renewable Portfolio Standard (RPS) Law* in 2003, and demanded that energy companies give a certain share of their total energy supply to renewable energies. Otherwise, they have to obtain green energy certificates on the market. Later, the Japanese government developed a "New National Energy Strategy" to enhance energy security. It's goal is to change Japan's conventional energy structure that depends heavily on oil. The government took 2030 as a target year, and set up several quantitative targets.

> Incentives are adopted by many countries in the early days of their PV development. The incentives include tax breaks, subsidies, interest subsidies, leasing, electricity quotas, and the like. However, the effect of those incentives is limited in many countries, with Japan as an exception. Due to their different political systems, it's hard for other countries to follow Japan in this regard and to maintain such a huge financial subsidy in the long term. On the one hand, government financing will be questioned in different quarters; on the other hand, changes in government or governmental officials will hamper follow-through on policy, leading to a lack of continuity in policies.

Unfortunately, after a wave of rapid growth in the PV industry, Japan's policies began to change. In 2005, the government stopped subsidizing solar power, with the excuse that solar power had been widely adopted by residents. As in other countries, once subsidies are reduced or removed, the development of the PV industry faced a tremendous impact. Following this move, Japan's installed capacity was exceeded by Germany in 2006, and its cell production was outdone by Europe in 2007. These figures show that Japan has lost first place in the solar power field.

Without question, in the decades before 2006, the strong support of the Japanese government, active follow-up from companies, and the participation of the whole nation played a fundamental role in the establishment of Japan's PV industry. After those years of effort, with the support of the "New Sunshine Program" and other follow-on incentives, an internationally competitive PV industry has formed in Japan.

Although the Japanese PV industry has grown slowly because of policy changes after 2006, vocal support for solar power can still be heard in Japan. The government has started to review policy changes, and "re-boost" solar power since 2009.

In 2009, the Japanese government began to discuss a policy that obliged Japanese electric utility operators to purchase electricity generated with renewables. This policy was adopted in 2011 by Congress and became the prototype of the later grants policy.

Compared with European countries and China, Japan invests more in its grants for solar power generation. The following policy on solar power subsidies was enacted on July 1, 2012. For non-residential projects of more than 10 kilowatts, the pre-tax subsidy reached 42 yen/kWh (about 2.6 yuan/kWh) for a period of 20 years. This amount doubles that of Germany or Italy, and triples that of China.[7]

Japan's huge investment in the PV industry is highly relevant to the 2011 Fukushima nuclear accident. After that, Japan became more determined in developing solar power, and announced a gradual shift to becoming a non-nuclear country. In the light of its domestic resources, it considered solar energy to be the best alternative to nuclear energy. In recent years, solar power has re-boomed in Japan. EPIA (European PV Industry Association) data shows that Japan's installed capacity of solar power amounted to 3,620 MW in 2010, 4,620 MW in 2011, and 6,000 MW in 2012, increasing by 30% over 2011.

Table 3–4 New PV capacity in Japan over recent years

Year	New PV installed capacity (MW)
2004	272
2005	290
2006	287
2007	210
2008	225
2009	483
2010	991
2011	1296
2012	1637

Source: Global New Energy Development Report (2013)

With large-scale development, the cost of solar production is declining, production is increasing, and the industry is booming. A number of leading enterprises have emerged. It's reported that Orix Corporation, a leading company in Japan's comprehensive leasing field, planned to cooperate with West Holdings, Japan's largest installer of solar power systems, and other companies in building 250 solar power plants in Japan, generating 500,000 kWh annually in total. The investment within 5 years is expected to reach up to 100 billion yen.

While global PV suppliers (including Chinese ones) in the solar power industry are being drawn to Japan's expanding market for a share, Japanese domestic firms continue to dominate due to geographic location and consumers' strong preference for domestic brands.

Distributors in Japan have established long-term cooperative relations with local brands, so it is difficult for Chinese PV companies to penetrate into Japan's closed PV market in the short term. Besides, Japanese customers are demanding in terms of the quality and safety of their PV products. Each package of imported PV products must be certificated before entering the Japanese market; otherwise, the entire shipment of modules will be refused. In other markets, customers just check whether suppliers have a German TÜV[8] mark; Japanese customers will test the products themselves. That means a higher standard and a more time-consuming process. Therefore, Chinese PV suppliers will find it isn't easy to find their footing in the Japanese market.

In addition, Japan's PV development structure is in line with the international trend of distributed generation. Its PV market is dominated by residential power generation, which accounted for more than 85% of its domestic market share in 2012, while that in Germany and the United States only took up 67% and 46% in their domestic markets, respectively.

Undoubtedly, it is the right choice for Japan to develop a market dominated by household electricity generation and consumption. First, it effectively promotes large-scale construction of rooftop generation facilities, reducing the power loss caused by feeding into grids, and improving the efficiency of power use. Second, it encourages end-users to collect electricity directly from their own devices, avoiding possible rising costs and duplicate charges after their electricity is absorbed by grids, and improving the popularity of rooftop solar power facilities.

Essentially, in the re-awakening PV industry, Japan has developed an ambitious plan with a series of incentives. Based on the good foundation it built before, Japan will have a less bumpy harvest in its PV industry. However, being subject to uncertainties, its road forward will still not be smooth. Now facing new competition from China and other emerging PV powers, Japan will have to solve such problems as its limited market, complacency, and shortage in innovation.

South Korea's Quiet Emergence

Like China and Japan, South Korea has also paid attention to the development of the world energy market. It has prioritized the development of renewables as part of its "national energy strategy," and supported them with preferential policies. Those moves catalyzed its soaring solar energy industry.

The South Korean government's supportive policies can be summarized as "support from government, investment from business, and participation by local communities." To be specific, they include general PV subsidy programs, a "One Million Green Homes" program, a regional installation subsidy program, a public buildings responsibility program, and a renewable obligation quota scheme.

The South Korean government alleviates the financial burden of PV equipment manufacturers, producers of parts and components, installers, and operators by providing long-term and low interest financing to them in their initial investment. Data released by its Ministry of Knowledge Economy shows that the Government's financial support for new and renewable energy technologies reached 199.4 billion won (about 1.1 billion yuan) in 2008. Annually, each research project on strategic technologies related to solar-power receives 10 billion won (about 55 million yuan). The longest duration of such support is five years. By doing this, breakthroughs can be made in core technologies which in turn can be applied as soon as possible.

South Korea is the first among Asian countries to propose and carry out a "feed-in tariff" method (compensation for feeding in). The government provides PV installers with 15-year fixed compensation for purchasing their electricity. The government will also subsidize the disparity between the standard prices published by the government and the purchasing prices at which solar stations sell their electricity to the state grid. For the 15 years starting from 2006, the Korean government has pledged to subsidize 0.56 to 0.6 euros/kWh, which is attractive to investors.

Stimulated by the feed-in tariff, South Korea's PV installation reached new heights at 20MW in 2006 and 2007. However, because the total amount of subsidies is limited and lasts only for a certain period, potential users hesitate to enter this sector.

In the first half of 2008, the South Korean government revised its tariff policies by removing the upper limit in total subsidies and resetting the tariff when the total applied installation exceeds 100MW. The lifting of such a ceiling and expectation about the gradual reduction in feed-in tariffs in 2009 boosted domestic installation demands. In 2008, its PV market witnessed explosive growth, ranking it the fourth largest country in terms of newly installed PV capacity.

According to EPIA statistics, by the end of 2011, South Korea's PV installed capacity totaled 750MW, ranking No.12 in the world, and third in Asia after Japan (4,910MW) and China (2,950MW).

In 2012, based on its new industry development plan, South Korea started to build new 100MW solar power facilities, and expected to put them into use in early 2013. The South Korean government's 2013 budget for PV research and development was about 198.8 billion won (about 1.1 billion yuan), increasing by 36.9% over the previous five years (from 2008 to 2012). The short-term goals of many research programs in South Korea are developing crystalline silicon (c-Si) solar cells with high efficiency, amorphous silicon thin-film solar cells and CIGS thin-film cells. The long-term goals and innovative objectives of most projects focus mainly on organic solar cells and dye-sensitized solar cells (DSSC).[10]

Although South Korea is a latecomer in the competitive PV market, its notable growth has positioned it in the worldwide lead in producing polycrystalline silicon, which is a main raw material in the PV industry. However, because the global PV industry is in a downturn recently, the South Korean PV industry is also caught in the overcapacity dilemma. Major companies have to cut or stop production, and even file for bankruptcy.

Meanwhile, in the second half of 2012, South Korea's polycrystalline silicon products suffered from anti-dumping investigations from China. In July 2013, China's Ministry of Commerce issued a lower tax rate than expected after its preliminary judgment, but this move exerted a negative impact on Korean companies, even leading to poor performance by some producers. Half of the polycrystalline silicon products exported from South Korea to China are provided by the OCI, a South Korean green energy and chemical supplier. It was reported that the OCI's Gunsan plant only used 50% of its production capacity, while Hankook Silicon Co., Ltd and Woongjin Polysilicon Co., Ltd. (ranking second and third respectively in sales to China) had filed for bankruptcy protection, and KCC's *Chungcheongnam-do* plant declared its intent to stop production after only one year of operation.

In response to obstacles in exporting, the South Korean Ministry of Knowledge Economy plans to develop its domestic demand in the short term to raise operating rate in its PV industry. To get the industry back on its feet, the government will build a total of 260MW of generation facilities for solar power in the next three years. RPS (Renewable Portfolio Standard) also has been raised from 230MW to 330MW in 2013. The government will provide tax rebates and other means to facilitate renewable energy export, and companies will find it easier to obtain export guarantees.

In the next five years, the South Korean Integrated R&D Center for Solar Projects will invest 150 billion won (about 830 million yuan) for the research and development of key PV technologies. If South Korea can successfully promote these projects, its sluggish solar industry will be revived.

Japan and South Korea Faltering

The development of Japan's PV industry maintains good momentum, but is still hampered by such factors as a limited market, complacency, no breakthroughs in core technology, and an incomplete industry chain.

Some believe that PV manufacturing in Japan cannot avoid the fate of its industrial development: leading the world in the early days, then being caught up by emerging countries in the middle term, and finally suffering a crushing defeat. One can find examples in semiconductor DRAM (dynamic random-access memory) chips, LCD TVs, DVD players, lithium batteries for digital household appliances, and car navigation devices, etc.

For example, both Sharp, the Japanese panel maker, and Kyocera, a maker of electronic components, have been developing solar PV technologies for 30-odd years, and their products have taken a substantial share in the global market. However, as China becomes a major cell producer, it, together with other emerging countries, is taking the place of Japan, whose global market share is diminishing.

How can emerging countries catch up with Japan so quickly? It is a question also asked by Japanese manufacturers, but is not too difficult to answer. The attrition of Japan's competitiveness is highly relevant to its closed mode of production, which can lower the risk of technology outflows, but likely push up production cost correspondingly.

In fact, it is difficult for Japan to protect its PV technology. Many Japanese manufacturers have to outsource their PV cell production to cut cost. So, even patent application is still insufficient for protecting their core technologies from flowing out via the production equipment. While it is difficult for Chinese manufacturers to obtain foreign PV production equipment, obtaining it will make it easier for China to mass-produce equipment, and shorten the production cycle.

As early as several years ago, leading Chinese PV cell manufacturers had adopted production equipment made by Germany and Japan. Core technological secrets can leak through via this equipment or processes, even though they were well kept by the two countries. The Chinese manufacturers tried to discover the core technology buried in the production equipment, and enhanced their technological competitiveness as a result.

In today's world, technological exchange has been enhanced, which will also shorten the time in which Japan can maintain superiority in the PV sector. Japan's share of DRAM chips in global markets dropped from 60% to 30% over eight years (from 1991 to 1999); its share of solar PV cells fell from 50% to 25 % in just two (from 2004 to 2006).

South Korea and Japan started their PV industry earlier, but grew complacent in their achievements. In addition, their domestic markets are small. Now, most Japanese and South Korean companies are managed by professional managers or the second or third generation of the people who set up these companies, without the pioneering spirit of entrepreneurs.

After being defeated by their international competition, Japanese companies hope to defend their domestic market share by making use of the strong preference for domestic brands of the Japanese people. But this strategy is flawed: Japan's domestic market is still vulnerable to penetration by its foreign counterparts; in addition, the Japanese market is simply not as large as that of China.

The Japanese government plans to install 28 GW of solar PV capacity by 2020, and 53GW by 2030. It's possible to apply 230GW in the long run. According to recent discussions, China has more ambitious targets than Japan in terms of installed capacity: 100GW in 2020, 400GW in 2030, and 1000GW in 2050. In light of China's vast territory, the application potential in the long term is immeasurable. Therefore, Chinese manufacturers have many more opportunities than their Japanese counterparts in terms of domestic market share.

In addition, Japanese PV manufacturers have been facing a bottleneck in making breakthroughs in core technologies, which makes it difficult to lower prices and raise competitiveness. In recent years, the conversion efficiency of crystalline silicon cells has almost reached its peak. Attempting to improve efficiency, Sharp cooperated with the Institute of Industrial Science, Tokyo University, to develop quantum solar PV cells. However, the institute also acknowledges that it will take at least 10 years to mass-produce their new cells.

With no significant breakthrough in technology, it's easy for Japanese makers to lose to their Chinese counterparts who can attract clients with lower prices. The penetration of Japanese manufacturers in the global market keeps declining. Moreover, they stick to a vertical integration model rather than a horizontal one. So it is likely for the entire solar PV industry there to repeat the mistake of flat-screen TVs. Given the context that most Chinese manufacturers are horizontally integrated, Panasonic, the Japanese electronics giant, announced in December 2011 that it would set up a Malaysian branch, and undertook vertical integration.[10]

Japan's stagnation in developing PV manufacturing technology is related to the absence of entrepreneurship, because most of the large companies are managed by professional managers.

Take Sharp for example. On June 24, 1980, its founder Tokuji Hayakawa on his deathbed was still worried about the company: "We should not fear imitation by our competitors. That shows we are competitive." Sharp had been an innovator since its birth. Although its products were often copied by others, it still led the trend with new products through continued innovation. However, in the 21st century, Sharp has slowed down its innovation. The high level managers just wanted to enjoy the fruit of its extant success, and to raise the barriers for getting access to its technologies. Departing from its founder's spirit of pioneering innovation, Sharp went downhill.

The Japanese solar PV industry is now at a critical juncture. If it is unable to win the competition with foreign businesses, a reshuffling will likely occur in this sector, further weakening its competitiveness. To put it simply, it is still unable to compete with the Chinese PV industry.

Besides suffering from similar challenges to those faced by Japan, the South Korean PV industry is also trying to get out of said difficulties. In 2010, the Samsung Group, a business giant in South Korea, entered this sector. Due to the limited domestic market, the price of cells plummeted after they were oversupplied. As a result, Samsung SDI has lost 30 billion won (about 165 million yuan) since it took over the PV cell business. Obviously, South Korea's PV market is still in its infancy, and there is still a long way to go. It is still not a strong competitor to China.

That doesn't mean Korean PV manufacturers have no chance to succeed. The anti-dumping and anti-subsidies duty investigations on Chinese-made solar panels by the US may do South Korea a favor. In fact, Chinese PV products account for 50% of the US market. Because of these investigations and others, Chinese companies will find it difficult to further penetrate the US market. But it will be a good opportunity for South Korea.

The Rise of the PV Industry in China

China's PV industry boasts more than 580 companies and 300,000 employees. The production capacities for silicon wafers, cells and modules have all reached

40GW, 60% of the total output of the world. After 10 years of effort, China has without a doubt become the world's biggest PV manufacturer.

However, China's PV industry is merely "big," rather than "strong." Problems such as exporting products made with imported raw materials, over-capacity, and an incomplete industry chain led to the industry-wide crisis in 2012. The sector needs to be integrated and upgraded while expanding to domestic demand.

As a big PV country, China has obvious competitive advantages but faces many challenges as well. Much effort is still needed.

The Concern of the Chinese Premier

When visiting a PV enterprise in Xingtai, Hebei Province, on June 7, 2013, Chinese Premier Li Keqiang asked the workers, "Considering the recent difficulties faced by the PV industry, are you able to tide over the trouble in the coming two months?"

This scene was widely reported by the media. In Hebei Province, Premier Li tried to pin down the effects of the EU's anti-dumping and anti-subsidies duty investigations on Chinese businesses. He assured the companies that, facing the complicated international situation, the Chinese government was resolute in its duty to safeguard national interests and oppose trade protectionism. He hoped that we would have the requisite confidence that we would pull through the temporary difficulties together.

Before his words of assurance, Premier Li himself had done a lot of critical work during his diplomatic trip in Europe. In late May 2013, he met with German Chancellor Angela Merkel and won her support. In the press conference, both of them expressed their positions against charging permanent tariffs on Chinese PV products.

After returning from Europe, Premier Li, in his call with EU President Jose Manuel Barroso on the evening of June 3, said that China would unswervingly fight against the EU's punitive duties. According to reports, it was the first time high-level Chinese officials clearly showed such an unswerving anti-sanction attitude towards the EU.

China's efforts bore fruit. On June 4, the European Commission decided at the last minute to impose provisional anti-dumping duties of 11.8%, down from 47.6%.

On June 14, Li chaired a State Council executive meeting, and announced six new policies to support the PV industry. On the same day, China New Energy Chamber of Commerce held an industry summit on new energies. At a time when they were worried about China's PV development, the participating companies were encouraged by Li's supportive policies.

In early August, China and the EU reached a "price guarantee" agreement after arduous negotiations. The EU decided to remove its anti-dumping duties on Chinese PV companies.[12]

Within a few months, the troubled Chinese PV players had enjoyed national support and the Premier's personal assistance, which had never happened before in the sector's history, and had not been seen in other sectors either.

For Primier Li, the PV industry is important in developing new energy and upgrading China's economy. For the State Council, this industry is internationally competitive, but faces operating difficulties for the moment.

That the PV industry enjoys such treatment in China is highly relevant to the industry's characteristics and achievements over the years. It started late, but has grown rapidly in the last decade. A complete industrial system has come into being, along with a whole industrial chain. It is one of the most competitive PV markets in the world.

When studying how China's PV industry rose, we need to understand the main development stages and related background of its history.

As early as the 1950s, PV in China was still wholly in the research stage and could not be put into operation. It is the 1980s that can be seen as the real start of this industry. Since then, there have been five main stages: infancy (before 2000), childhood (from 2002 to 2007), setback (from 2008 to 2009), recovery (from 2009 to 2010), and turmoil (from 2011 until now).

China's application of solar PV began in the 1970s, but it wasn't until 1983 that it saw real growth. From 1983 to 1987, seven production lines were introduced from the United States, Canada and other countries. China's annual production capacity of solar cells jumped to 2.1MW in 1998 from 200kW (before 1984). The cumulative production was more than 13MW with a cost of more than 2.5 yuan/kWh.

In 2002, China launched a program with the aim of providing PV generators and small wind turbines to seven western provinces, especially townships there that had no access to electricity. This program spurred the PV industry. Packaging lines for solar cells were set up in China, with cell production soaring year on year. In October 2003, the National Development and Reform Commission (NDRC) and the Ministry of Science and Technology formulated a five-year plan to develop solar energy. The NDRC also provided 10 billion yuan to the "Brightness Program" to promote the application of solar generation technologies. By 2005, the total installed capacity of solar systems in China reached 300MW. In 2007, China surpassed Germany as the world's number 1 cell producer.

Over the following two years, the global financial crisis broke out. China's PV industry dropped to rock bottom for the first time after encountering related setbacks. Many Chinese enterprises were hit seriously, but the industry was still promising in the eyes of most people. After 2009, supported by national policies, China's domestic PV market recovered to a certain degree. In 2010, PV orders increased substantially, and the market began reviving.

In 2011, the first year of the 12th Five-Year Plan (2011–2015) in China, the PV industry chain was improved, new solar projects were launched nationwide, and the sector welcomed a golden age, thanks to supportive policies.

However, the world economy deteriorated dramatically. From 2011, a global downturn was to be seen everywhere. In 2012, the anti-dumping and anti-subsidies duty investigations carried out by the United States and the EU exerted a negative influence on the Chinese PV sector, hitting PV production, management and sales. Solar enterprises like Suntech and LDK were troubled by serious debt crises.

In light of those difficulties, the State Council issued the *Opinions on Promoting the Sound Development of the PV Industry* in July 2013. China's PV industry would see a new round of rapid growth.

According to the target set in the *Opinions*, expanding the domestic market, improving technology, and speeding up industrial restructuring and upgrading are the ways to ensure the sustained and healthy development of the PV industry. China should set up its PV system to cover production, sales and services. Regulations, policies, standards and market should be formed to favor development. From 2013 to 2015, the newly installed capacity each year will reach 10,000MW and be above 35,000MW by 2015. China should accelerate the mergers and acquisitions of companies, shut down those with low product quality and obsolete technologies, and cultivate leading players that are strong in R&D and advantageous in market competition. China needs to significantly reduce the cost of PV generation by accelerating technological innovation and industrial upgrading, as well as improving self-sufficiency in raw materials such as polycrystalline silicon and cell manufacturing technology. Chinese products should occupy a reasonable place in the international PV market and China should make new progress in foreign trade, investment, and financial cooperation.

"A Superpower in PV Manufacturing"

China is undoubtedly a superpower in manufacturing PV in the world today.

The manufacturing chain for crystalline silicon PV is composed of silicon materials, wafers, cells, modules and systems. In recent years, China's PV industry has grown rapidly. The scale of cell manufacturing has expanded dramatically, giving China a large share in the world market and world-class technology in the field of cell manufacturing. Polycrystalline silicon smelting technology has matured. China has established a complete industrial system including silicon materials, wafers, cells, modules, inverters, and controlling equipment. The domestic application market is booming, with costs reduce significantly. All this proves that China is becoming more competitive.

Rapid development is also seen in crystalline silicon manufacturing equipment and supporting industries. China has become the world's largest producer of wafers (including monocrystalline wafers and multicrystalline wafers). Wafer manufacturers are mainly located in Liaoning, Hebei, Henan, Jiangsu, Zhejiang and Jiangxi provinces; cell and module manufactures are mainly based in Jiangsu, Hebei, and Zhejiang, with their collective output ranking first in the world. The output of polycrystalline silicon and inverters is relatively low. At this point, the sector has seen more than 50 companies starting to make polycrystalline silicon, for which China is likely to be an exporter in the future.

So far, China is home to more than 580 PV companies with 300,000 employees. In 2007, domestic cell production was about 1180MW, nearly a threefold increase over the 300MW of 2006. Moreover, in the same year, China surpassed Germany as the world's largest cell producer. In 2011, China's production capacity for wafers, cells and modules reached 40GW, a 100% increase compared to that of 2010 and accounting for 60% of total global output. China has become a real superpower and later a center for PV manufacturing, with production increasing still further.

In 2012, China's production capacity for polycrystalline silicon reached 158,000 tons, accounting for 43% of the global total. The output was 69,000 tons, 32% of the world aggregate. China's production capacity of PV modules was 37GW, 51% of global production, while its output of modules was 22GW, representing 54% of the world's total.

China's polycrystalline silicon output is 143 times higher than that ten years ago. In the past, advanced technologies were mainly in the hands of the United States, Japan, and Germany. To be specific, they were held by seven companies and ten factories, which monopolized PV technologies and blocked China from introducing them. In 2002, the global output of polycrystalline silicon was 20,350 tons. China only produced 50 tons, 0.25% of that total, along with its low level of technology, small production volume, high energy consumption per unit of product, and high production cost. Because almost all the demand for polycrystalline silicon produced in China came from abroad, the development of new energy was restricted by other countries.

2007 was the first year China mass-produced polycrystalline silicon. With its own research and technology, SINOSICO, based in Luoyang (capital city of

Henan Province), put into operation an expansion project with a production capacity of over 1,000 tons. Sichuan Xinguang Silicon Technology Company put its project into production with an annual capacity of 1,000 tons. By the end of the year, the national output of polycrystalline silicon had reached 1,130 tons, an increase of 295%. China's large-scale production of polycrystalline silicon had started.

In 2008, driven by the rapid growth of the industry and high profits, there was a wave of polycrystalline silicon building projects in China. More than 20 projects were under construction, with a total investment of 20 billion yuan. A large number of new and expanded projects had been put into operation, including the GCL project in Xuzhou, the third phase of SINOSICO, Emei Semiconductor Material Factory, Chongqing Daqo New Energy Corp., Jiangsu ShundaSolar, Asia Silicon Qinghai Co., Yichang CSG Polysilicon Co., Ltd., Tongwei Co., Ltd., TBEA Co., Ltd., Ningxia Sunshine Silicon Industry Co., LDK, and Zhongcai Group. The cumulative production volume totaled 20,000 tons. In that year, China's production of polycrystalline silicon reached 4,500 tons, an increase of 298%. But more than 70% of the product being used in China was imported. Its price hit USD 500/kg, a record high, which caused capital to flow in.

Shaken by the financial crisis in 2009, this price plummeted, while China's total production capacity climbed further to 80,000 tons, producing 20,230 tons that year. In 2010, the production scale hit 120,000 tons, with an output of 40,000 tons. In 2011, the production scale was 165,000 tons, with an output of 82,670 tons, accounting for 35.9% of total global output (230,000 tons). China for the first time surpassed the United States as the world's largest producer of polycrystalline silicon.

The polycrystalline silicon sector is just one link in the PV industry chain. Taking a close look at the entire value chain of the PV industry in China, we find that the numbers of PV suppliers in each link are in a pyramid-shaped distribution.

On one hand, there are fewer producers of polycrystalline silicon upstream, and they need a longer time to increase production. In the middle and lower stream, there are more cell and module manufacturers who need less time to expand. In the whole value chain, the expansion of crystalline silicon manufacturing lags far behind the production of cells and modules.

On the other hand, generation devices with crystalline silicon are different from conventional ones. It's also easier to make and maintain them. Analysis shows that the technologies for producing silicon wafers, cells and modules downstream are relatively mature. In recent years, domestic enterprises have made significant progress. Many of them have become world-renowned at making crystalline silicon cells.

The rapid growth in all links has guaranteed the expansion of China's PV industry and given it an edge in international competitiveness. However, China is a relative newcomer and, after a decade of growth, suffers from serious over-capacity because its domestic market is yet to be tapped. It was also hit by the aforementioned investigations imposed by the United State and the EU. Now, this industry in China has to face a fresh round of turmoil.

The major repercussions for China's PV industry came from the uncurbed expansion of the global PV industry for two consecutive years (from 2010 to 2011). At the end of 2011, global PV production capacity was over 79GW, and the total output was 40GW, compared with an installed capacity of only 27.7GW at that time. The serious imbalance between supply and demand led to dropping module prices from 2011 onwards. The price slumped from USD 0.9/watt at the beginning of 2012 to USD 0.65/watt at the end of that year. With corporate profits plunging, it was not uncommon for PV companies to apply for bankruptcy. After the reshuffling in 2012, global PV capacity diminished to 70GW, and output to 39GW, still higher than the newly installed capacity of 31GW that year.

This marks China's transformation as a superpower in PV manufacturing and indicates that further integration in its PV industry is underway.

Transitioning to a Big PV Consumer

The golden opportunity for transitioning appeared in 2012. Both the United States and the EU had imposed anti-dumping and anti-subsidies duty investigations on solar products exported from China. Their move posed tough challenges for China's PV sector. It had to consolidate the industry, improve technology, and develop a market for it. Among those measures, expanding the market is considered the most effective.

Previously, China's PV industry had to import raw materials and export products. Its small domestic market had seriously hindered the attempts of PV technologies to advance. At the same time, international demand, exchange rates, and foreign subsidies have a direct influence on Chinese PV makers. Therefore, the best way out is to expand domestic applications by developing the market for distributed PV generation and carefully constructing PV power plants, and so on.

In fact, it's obvious that the domestic solar power market has opened up on a massive scale in recent years. In 2012, the newly installed capacity registered 4.7GW, increasing by 80.3% year by year. The cumulative installed capacity reached 8.2GW. With the development of a distributed generation market, PV technologies will be applied in a faster manner. Supported by national policies, China's new PV installations are expected to reach 10GW in 2013.

In the future, the newly installed capacity in Germany and Italy will increase at a much slower pace, while the PV markets of China, the United States, and Japan will be more prosperous. It is expected that the latter three countries are going to generate 47% of the total demand for PV in the world in 2013. Thanks to policy incentives and an optimized industry chain, China is going to replace Germany as the world's largest PV market.

The first impetus necessary for a domestic PV market comes from the "solar rooftops" which are fairly common nationwide. By launching the latest incentives, the State Council encourages electricity consumers to build distributed power generation systems, so that they can meet their own demand for power first, and inject the surplus into grids which in turn can be transmitted to those in need. Industrial and commercial businesses and industrial parks were given top priority in building large-scale systems, because they had to pay a higher price for electricity before. It also supports small-scale ones to be set up in schools, hospitals, government agencies, institutions, community buildings and other such buildings.

Second, urbanization will be a growth point. During this process, China should give solar energy a chance to thrive by encouraging more people to use PV generation, for example, to build energy-efficient buildings and promote BIPV. For example, such applications can begin with demonstration cities for new energies, demonstration counties for green energies, and demonstration cities (or counties) for renewables, and can be spread to more districts, towns and villages of this kind.

Urbanization will be the new engine for China's economic growth, and also for PV generation. In the stage of planning urban power supply, we can introduce this technology as a supplement. For example, the rooftops of public buildings and new residential houses are good places to develop BIPV applications.

Appropriate policies are also essential to stimulate domestic demand. The State Council has issued relevant guidelines to encourage pilot and demonstration programs for the micro smart grid for new energies, in which distributed generation can be applied and used on a large scale. It also encourages the development of power management and operational mechanisms, the designing of new models of construction, the operation and consumption of distributed PV generation, the support for PV power generation in remote areas and islands by offering them access to power supply, and the encouraging of distributed PV lighting in urban streets, public areas, base stations of telecommunications, and traffic lights.

In addition, local governments will drive the construction of PV power plants in an orderly manner. They hope to arrange PV generation along an appropriate geographic distribution, help the stations get access to the nearest grids, encourage PV electricity to be consumed in the local community, and build more stations with good planning. By considering the growth of their power market and the need for restructuring of the energy sector, local governments will carefully push forward the construction of PV plants in line with the consumption potential of the market. They will encourage existing grid facilities to build PV plants, which will grow side by side with the traditional ones. The construction of PV plants should be at the same pace as the planning and construction of supporting grids, to ensure timely connections and efficient applications.

The establishment and implementation of these policies will be good news for PV generation in China. It will digest the surplus capacity in PV manufacturing, and facilitate China's transformation from a big PV producer to a heavy consumer. Solar energy, the most promising among the new energies, will boom on a massive scale in China, and finally become a substitute for, rather than a complement to, fossil fuels.

Indeed, China has great potential in this regard. Statistics show that the construction area of buildings in China is nearly 50 billion square meters, which will get close to 90 billion in 2020. If we can install solar cells (with a conversion efficiency of 10%) on 10% of rooftops and 15% of outer walls, the total installed capacity will reach 1,000GW. Given that the acreage of decertified land in China approched 2.62 million square kilometers, we can install 2,000GW (calculated as 0.04GW/m²) of PV facilities in just 2% of it, twice the installed capacity of power supply in China now.

Table 3–5 China's newly added PV capacity over time

Year	Newly added PV capacity (MW)
2004	9
2005	4
2006	12
2007	20
2008	45
2009	228
2010	520
2011	2 607
2012	4 700

Source: Global New Energy Development Report (2013)

Undoubtedly, the increase in subsidies will push the domestic market to open up. Currently, according to financial sources, the subsidies fall into two categories.

Some of them are from the Ministry of Finance (MOF). According to its *Opinions on Accelerating the Application of Solar PV Construction* (released in March 2009) and *Interim Measures for the Management of Financial Subsidies of Golden Sun Demonstration Projects* (launched in July 2009), the government will subsidize BIPV projects and those specified in the appendices of the Golden Sun Projects. If a project is approved, the subsidy will be granted in its installation stage in a lump sum. In July 2013, the MOF released the *Notice on Subsidies to Distributed Solar PV Power Generation according to Power Output*, clarifying that the state will subsidize distributed PV power generation by power output, and grid companies will give the subsidies directly to PV plants. This policy will give distributed PV plants new impetus.

Others subsides go to additional supporting projects for renewables. After a price tender, the government grants a subsidy to the winners according to a benchmark price. The beneficiaries of the Golden Sun Demonstration Project would be excluded from this subsidy. Therefore, we can see that China's PV market is supported by the two groups of subsidies, and that they are independent of each other.

Over the last few years, as PV manufacturing capacity has expanded rapidly, manufacturing technology has become more sophisticated, improving the productivity and management process by a large margin. The cost and price of polycrystalline silicon, modules and inverters have dropped dramatically. At the same time, by constructing and maintaining large-scale PV plants, the industry has gained considerable invaluable experience in plant design and construction and operation and maintainence, which in turn lowers the cost of PV power plants.

By 2012, the investment in large-scale ground PV plants had reached 1.5–1.8 dollars/watt on average in China, and the cost of PV generation had decreased correspondingly. In sunlight-rich areas, the generation cost of such big plants has dropped to about 0.6 yuan/kWh. This is the most tangible benefit brought to the Chinese and people of the world, now more confident that PV will replace traditional energies one day in the future.

As the downstream PV industry is now more profitable, more and more upstream manufacturers have begun to build plants. According to the Worldwide PV Manufacturing Database maintained by SEMI, the top 20 PV manufacturers of modules and cells in China have also been involved in the plant development business. In 2012, China's investment in PV plants totaled 45 billion yuan, and it was expected to reach 100 billion yuan in 2015. Undoubtedly, the market potential will be huge.

Since the Advantages are Obvious, More Effort is Needed

The year 2012 was crucial both for China's new energy industry, and for its PV industry in particular.

That year China ranked high in the new energy sector in the world. With investment up to USD 67.7 billion, 25% of the global total, China, followed by the United States, had become the biggest investor in this field. Most of its investment goes to the PV sub-sector.

In that year, the difficulties encountered by the Chinese PV industry were also opportunities for restructuring and upgrading. In particular, PV generation costs dropped dramatically. That will help the expansion of its domestic market, and strengthen the competitiveness of this industry in China.

Chinese PV products frequently suffer from anti-dumping and anti-subsidies duty investigations in the international market. This demonstrates exactly the competitive strength of these products. Compared with their European counterparts who have received massive subsidies, Chinese companies are competitive in price. To put it simply, China's PV exports are hindered because of trade wars frequently waged by the EU and the US. It was not their quality that lead them into that difficult position.

In recent years, it has been internationally recognized that China has the manufacturing technology advantage in regard to silicon wafers, modules and PV systems. Its technologies have reached or already surpassed the international standard. In addition, its energy prices and labor costs are lower. Therefore, China's PV industry has built up strength along the entire industrial chain. Boasting low cost, high efficiency, advanced technologies and reliable quality, China's PV products have gained popularity among major users and ordinary consumers in the US and Europe. At present, China's solar energy industry has gained the world's attention and its production capacity for cells has surpassed that of Europe and Japan. We have established a complete industrial chain, from producing raw materials to PV generation.

Although China has obvious competitive advantages, its PV companies need to put in yet more effort. If an emerging industry wants to develop rapidly, the related sectors upstream and downstream need to develop in a synchronous manner. The development of an industry does not simply mean increasing its production capacity. More importantly, it must enhance international competitiveness with its own core technologies. Because the US and the EU imposed investigations, and PV applications are slow in the international market, China's PV industry has suddenly reached an impasse. This indicates that it is prone to risks, and its roadmap for polycrystalline silicon cells is defective.

In the first half of 2013, China's polycrystalline silicon industry faced complicated situations at home and abroad. Companies were in a dilemma. In essence, the downstream demand declined catastrophically, foreign dumping continued, prices remained low, and hopes to resume production were faint.

Demand was reduced by the investigations. In the second quarter of 2013, China's export of PV products was hindered, and the demand for polycrystalline silicon fell to 69,000 tons, a year-on-year decrease of 4.2%. By contrast, from January to May 2013, polycrystalline silicon imported by China totaled 34,000 tons, an annual increase of 0.91%. In those five months, the import price of polycrystalline silicon averaged USD18.33/kg, dropping by 27.7% if compared with the full-year average import price of 2012. This suggests that foreign dumping was a present issue.

Affected by foreign dumping and the sluggish downstream demand, the prices of PV devices lingered at a low level. In the first half of 2013, the average spot price of polycrystalline silicon in China dropped to 130,000 yuan/ton, with a year-on-year decrease of 32.5%. Until now, its spot price is still lower than the production cost in China. If there are no favorable signals, Chinese PV manufacturers will see no hope of resuming production. It is predicted that the crystalline silicon PV industry will still face the problem of overcapacity in 2014. With the expansion of the global PV market and lower cost of PV generation, PV supply and demand will reach equilibrium after 2015.

The EU's Investigation Conspiracy

The parallel investigations imposed by the EU against Chinese PV products have hurt Chinese enterprises for two years. At last, thanks to its decisive response, the Chinese government has settled the disputes with the EU through the mediation of high-level officials on both sides.

However, we should note that many EU member states are cutting their governmental subsidies for native companies, while they impose investigations on Chinese PV companies. This is not simply due to the European debt crisis.

Settling the Disputes

November 7, 2011 was a day when Chinese companies in the crystalline silicon industry felt worried.

On that day, the United States revealed its final determination to punish crystalline silicon PV products coming from China. On the same day, the EU also announced its anti-subsidies duty investigations on the same products made by China. It was the prelude to more European investigations against China. Moreover, the EU's move was more lethal than that of the US, because the US only represents 10%, while the EU accounts for more than 60% of the total number of PV products China exported.

Thankfully, after nearly two years of investigations and negotiations, the result was mitigated. On August 6, 2013, China and the EU reached an agreement on price undertaking. China agreed to raise the minimum prices of its PV exports to the EU, and limited the total volume of annual PV exports to the latter. The EU in turn agreed to remove its anti-dumping duties of 47.6% on the Chinese PV exporters that adhered to the agreement.

According to relevant EU laws and to prevent price manipulation, the two sides agreed not to release the specific minimum price and annual export quotas. The price undertaking will be honored until the end of 2015. However, Bloomberg reported that the minimum price was 0.56 euros/watt, and the annual export quota of solar panels would not exceed 7GW.

The undertaking entered into force on August 6, 2013. Most Chinese crystalline silicon PV companies participated in the negotiations, and they did not need to pay the punitive tariffs from that date onwards. But those who did not are not be spared. Besides, Chinese PV companies are still subject to the punitive tariffs if their exporting volume exceeds the quota.

The PV trade dispute between the EU and China was finally settled. After overcoming this challenge and easing the temporary tension, Chinese PV companies still account for 60% of EU market share. However, the minimum price will make Chinese PV products less competitive, and China will lose part of its market as a result. The interim agreement also shows that the EU–China PV trade disputes have not been fundamentally resolved. When the agreement expires, the two sides need to negotiate again.

> It has been move than fifteen years since the "Renewable Energy Strategy" was formulated in 1997. Over those years, leaders have changed in the EU and other countries. Facing strong opposition and resistance from fossil energy companies at different times, foreign governments still unswervingly support clean energy. Their attitude towards clean energy is worth learning from.

In the PV revolution, Europe responded so quickly and effectively that the clean energy industry, in which solar energy is dominant, has become one of the core industries supporting the EU's strategic values. Undoubtedly, all forces that benefit solar energy deserve great support. However, in recent years, the EU has begun to cut subsidies for PV enterprises, and brings cases against China. How do we interpret those moves, which seem to be anomalous?

Besides the Sino-US and Sino-EU trade disputes, India began to launch anti-dumping and anti-subsidy investigations against Chinese PV products in 2012. China also imposed such investigations on the polycrystalline PV products imported from Europe, the United States and South Korea. The PV trade wars have spread worldwide.

Taking a look at the history and the status quo of international trade, we find that these wars broke out during the economic crisis, but that, more importantly, all major countries in the world have treated the PV industry as a key strategic industry for their energy security. They have done their best to make a forward-looking arrangement for it. Therefore, China's PV players are more rational in looking at trade frictions and have more confidence in the PV industry.

Global PV Trade War

The double investigations against Chinese crystalline silicon PV products were applied first by SolarWorld, the largest PV company in Germany. On October 19, 2011 and July 24, 2012, it led the double investigations in the US and the EU against China.

SolarWorld opposed the Chinese government's subsidies for PV products. It did not support the German government in cutting subsidies. The European Commission was willing to accept a price undertaking rather than a high anti-dumping duty at the last minute, largely because the German government was against such a duty. In the past five years, Germany has benefited from its increasingly close relations with China, and it doesn't want to destroy this relationship, even when facing pressure from major PV companies.

In fact, the Sino-European PV dispute highlights the complicated relationships between the European Union, Germany and China. The EU Treaty does authorize the Commission to investigate trade disputes, levy tariffs and conduct negotiations, but many relevant decisions must be approved by all EU member states. China can take this as an opportunity and negotiate a way out with each member state bilaterally.

China has taken a series of actions that are seen by many people as exerting pressure on European governments and forcing them to oppose the duty. China launched the anti-dumping investigation on French wine, and planned to levy a tariff on polycrystalline products imported from the EU. These two moves seriously undermined the interests of the German industrial giant WackerAG. China also claimed it would launch an anti-dumping investigation on EU-produced limousines. This may mean German auto giants such as Daimler AG, BMW and Volkswagen may be denied the chance to access the promising Chinese market.

However, China's investigations are not about winning the PV trade war, but settling the China-EU PV disputes.

In fact, before China launched them against the EU, it deliberately postponed announcing its preliminary moves against European polycrystalline, and only launched anti-dumping investigations against South Korea and the United States. This not only sent a warning to Europe, but also gave China time and space to settle its PV disputes with the EU. If the EU had introduced a high tax, there would be losses for both sides. The anti-dumping tax would push up the prices of the Chinese PV products, result in a lower installed capacity, and severely restrict the development of related industries.

The dispute also indicates the complexity of global manufacturing and the uselessness of protectionism. Anti-dumping duties will jeopardize the interests of the manufacturers of solar panels with crystalline silicon in China, increase the price of Chinese panels, and cause damage to Europe. The most immediate victims will be the consumers and fitting producers who purchase PV products for cheaper, greener energy. By contrast, anti-dumping duties will indirectly reduce generation capacity, and the consequences will be borne by all power consumers. The EU member states eventually realized that high anti-dumping duties would make the cost greater than the economic benefits gained by the protected companies. Besides, environmental protection is another relevant factor in this regard. Obviously, what the world needs are cheap renewable energies as opposed to merely protecting products made in Europe.[12]

However, unlike the settlement of the EU-China disputes on PV trade, China and the US failed to reach consensus on their disputes. The two sides are in the vicious cycle of mutual retaliation.

In October 2011, seven US-based cell producers applied to the US government for anti-dumping and anti-subsidies duty investigations against crystalline silicon PV cells imported from China. In March 2012 and May of that year, the US Department of Commerce preliminarily decided to levy a anti-subsidies duty of 2.9%~4.73% and anti-dumping duties of 249.96%~31.14% on Chinese cells. In October, the US finally ruled that China was dumping and subsidizing its crystalline silicon PV cells, and decided to levy tariffs of 34%~47% on most crystalline silicon panels and cells imported from China.

The high tariffs led to a substantial reduction in crystalline silicon PV modules exported from China to the United States in 2012. In August 2012, China's revenue from exports to the United States was only USD 85 million, dropping 80% compared with the USD 387 million in January.

In response, the Chinese government imposed anti-dumping and anti-subsidies duty investigations on US polycrystalline PV products. The MOFCOM decided to carry out provisionary anti-dumping measures against the United States and South Korea by collecting security deposits, which went into effect on July 24, 2013. The preliminary ruling said that the two countries were dumping their solar-grade polysilicon in China, and causing material injury to China's polysilicon industry. The dumping margin of the US ranged from 53.3% to 57%, and the South Korean from 2.4% to 48.7%. China imposed an anti-dumping duty of 53.7% on MEMC, the second-largest poly-crystalline maker in the US and Hemlock, a famous polycrystalline enterprise, and a duty of only 2.4% on OCI Company, the largest polycrystalline maker in South Korea. Obviously, the US makers paid much more than the South Korean ones.

In 2012, countries, including the United States, the European member states and India conducted anti-dumping and anti-subsidy investigations against China and levied duties accordingly. A trade war on green energy began. By 2013, those fights were no longer only against China. There were also frictions between the US and India.

Two months after the United States called for sanctions against India's renew-able energy sector in the WTO, the Indian government also made a similar move. India told the WTO that the US federal government and its States were subsi-dizing their local renewable enterprises, creating formidable barriers to India's exports, and undermining global trade rules. In February 2013, the United States resorted to relying on the WTO to settle its dispute with India. It said that under the Jawaharlal Nehru National Solar Mission (also known as the National Solar Mission), India was subsidizing its domestic solar cell and module manufacturers, and affecting PV production and business operations in the United States negatively. There were similar arguments between Japan and Australia.

The Underlying "Thin-Film Plot"

Tracing the evolution of the global trade wars in PV, it is no surprise that it was the European countries, particularly Germany, that were the first to launch their investigations against China. Germany, the most powerful voice in the PV industry, has few players engaged in the R&D and production of polycrystalline silicon cells and monocrystalline silicon solar cells.

So, why did Germany spearhead those investigations against China? There may be an underlying "thin-film plot."

Currently, the most advanced technologies in the PV industry are those related to crystalline silicon cells and thin-film solar cells. The former includes monocrystalline silicon cells and polycrystalline silicon cells. The latter covers amorphous silicon cells, cadmium telluride (CdTe) solar cells, copper indium gallium selenide (CIGS) solar cells, and the like. The majority of China's PV enterprises are producing crystalline silicon. The voices in support of the EU's investigations mainly come from the manufacturers of crystalline-silicon based solar cells.

Both Germany and the EU want to push up the price of crystalline silicon solar cells, and create enough market space for thin-film cells. After all, compared to crystalline silicon cells, especially those made in China, German thin-film cells are uncompetitive given their higher production costs.

Data in the German Industry Almanac shows that in the past five years, more than 60% of the German government's financial assistance for this sector goes to thin-film cells, and more than 70% of research funding focuses on thin-film. In addition, all German thin-film PV companies are entitled to subsidies, but crystalline silicon enterprises do not enjoy such favorable treatment.

In addition to electricity subsidies, thin-film PV enterprises also enjoy subsidized loans. In September 2011, the German government restored the subsidies for banking loans given to amorphous silicon thin-film cells, CIGS thin-film cells and CdTe solar cells with efficiencies of above 11.6%, 13.8%, and 15%.

All those moves fully illustrate that the German solar crystalline silicon businesses had started to decline. Because the efficiency of thin-film modules had been boosted in the past five years, the cell industry associated with it is widely favored and had attracted a lot of investment.

> The European Commission levies tariffs on Chinese crystalline silicon solar cells and pushed up the price, which might appear to be good for the "interests" of the European cell enterprises. Actually, this will by no means enhance their competitiveness. Instead, it does a favor to another branch of solar power technology—thin-film cells—in allowing it to get a bigger market share. It contributes to the development of emerging thin-film cell technologies, and ensures European leadership in the next generation of thin-film PV. The crystalline silicon solar cell manufacturers in Europe will be the victims of the EU's double investigations.

This is the underlying reason why the EU launched the double investigations.

Besides the investigations, there is also a misunderstanding about the cutting of subsidies to PV companies by EU member states. It is believed that they were forced by the European debt crisis to cut subsidies for solar energy, intensifying the competition between the European enterprises and their Chinese counterparts, and that this is what led to the investigations.

The European debt crisis did in fact have a negative impact on the solar industry there. However, it is inevitable for the products based on crystalline silicon to lose when thin film has come into the mainstream. European enterprises are not only encountering severe competition, but also technological difficulties which challenge all firms making crystalline silicon products.

All in all, the European Commission's cuts to the subsidies to crystalline silicon cells and its support for the more promising thin-film technology are conducive to emerging and competitive enterprises there. The fight against Chinese polycrystalline products is both a blow to competitors and protection for local thin-film products. The European Commission's investigations against China are deliberately planned.

The Era of Thin-Film PV is Here

Adopting thin film and flexibility are the future trends for the PV industry. The difference between crystalline silicon products and thin-film ones is comparable to that between Intel286 or Intel386[13] and laptops or tablets. It is safe to say that the thin-film PV era has arrived.

Five Advantages of Thin-Film Modules

Thin film is the future of PV power generation. In particular, by offering higher efficiency and flexibility, thin-film modules will be dominant in the future PV market.

There is a misunderstanding about solar power generation in China: it is believed that PV development only needs crystalline silicon. On the contrary, thin film is the new mainstream technology in global PV development.

> Thin film and flexibility are the general trends in solar energy development.
> First, the advantages of thin film include zero pollution, low energy consumption, and wide application. Thin film shows its superiority in low temperature coefficiency and good weak light performance. In addition, thin-film cell technology still has enormous potential to improve and to develop.

Second, thin-film cells are flexible and lightweight, which enable them to be applied widely, such as in solar emergency lights, street lamps, camping kits, chargers, and especially solar cars, which are most promising in this regard.

China has treated the PV industry as its strategic focus among emerging industries. "Emerging" means extremely promising. "Strategic" means it is important for the general and long-term interests of the country. So we must understand the trends; look into the future; seize the industrial opportunities;

provide political and financial support to R&D, applications of thin film, and improving flexibility; and welcome the era of thin film.

Q-Cells, the world's largest solar giant and the former leader in the polycrystalline sector, has gone bankrupt, which proves that crystalline silicon cells will be replaced by thin-film ones. What is the difference between thin-film and crystalline silicon? Drawing an analogy: if crystalline silicon were a black and white television, thin film would be an LCD TV.

Crystalline silicon faces many challenges: an extremely small market, high energy consumption, and high levels of pollution in its production. China is powerful in solar energy production, but weak in solar application. This means China produces the dirty and energy-intensive crystalline silicon at home, and supplies clean energy abroad. In addition, the production of crystalline silicon is complex, which makes production costs higher than thin film.

> Simply focusing on the conversion rate is a wrong move. Production costs, efficiency, and performance–price ratio are the most important factors to focus on. When the rising cost of thermal power and the declining cost of solar energy are equal to each other, the former will be replaced. Higher conversion efficiency is not always a good thing. There should also be an optimal performance–price ratio, in which the key is the cost per watt and the cost per kWh.

Comparing the properties, production processes, and applications of crystalline silicon and thin film, I think the latter has at least five advantages.

First, they consume fewer raw materials. The thickness of the wafer ranges from 150 to 200 microns, while the photosensitive layer of a thin-film cell is only 2 microns. Crystalline silicon cells consume 100 times as much raw silicon as thin-film cells do.

Second, they have a shorter payback period. Thin-film modules consume much less energy per watt than crystalline silicon modules. The cost of crystalline silicon is 1.5kWh/watt, while that of the thin-film modules can be set to be no more than 1kWh/watt. Therefore, the energy payback period of crystalline silicon modules is more than 1 year, while that for thin-film modules is only half a year.

Third, they have a smaller temperature coefficient and good weak-light performance. It has been shown that when the operating temperature is above 25°C, an increase of 1°C will make the maximum power of silicon-based thin-film modules, monocrystalline modules and polycrystalline modules reduce by 0.21%, 0.47% and 0.46%, respectively. Compared with crystalline silicon modules, thin film modules can be deployed in more regions. Since the temperature coefficient of thin-film modules is lower, its electric power is not easily affected by temperature changes. Within latitudes below 40ºC, with the same installed capacity, the electricity output of thin-film modules is 10% higher than that of crystalline silicon modules.

The fact that thin-film modules can perform well in weak light ensures power generation under a variety of conditions. Scattering will happen when sunlight meets the air, dust and clouds, especially in cloudy or windy weather, in the morning or at dusk. This property enables thin-film modules to work even in dim light. Besides, even in shade, the loss of its electric power is small. In practice, it is unavoidable for the BIPV to be in shade, and the power loss of thin-film modules is much lower than that of crystalline silicon modules under this inevitable circumstance.

Fourth, thin films can be made into flexible modules. With film being deposited on flexible substrata, the modules can be bent or stuck to the surfaces of buildings, and this is promising, especially in the BIPV market. On the one hand, flexible thin-film modules are lighter, and they can be installed on non-loadbearing roofs. And on the other hand, thin-film modules are flexible enough to be stuck to a variety of surfaces, and so can meet the structural and aesthetic requirements of buildings. Most of today's industrial plants use light-weight steel roofs, while traditional PV modules are rigid and heavy in weight, so they are not commercially available for installation on those roofs. Flexible thin-film modules are lighter and can meet such architectural requirements. Therefore, the existing massive rooftops in industrial plants are a blue ocean for this new technology. In addition, thin-film modules can be used in the auto industry, such as on sunroofs, vehicle roofs, and in flexible and portable devices for power generation in vehicles.

Fifth, diversified products. Compared with crystalline silicon modules, which have a monotonous product mix, thin-film modules can achieve diversified applications in many fields. In BIPV applications, thin-film cells can be made into transparent modules with adjustable hues. Thin-film modules can generate power without damaging the beauty and integrity of the host buildings, so they are the best choice. Since thin film is flexible, collapsible and portable chargers are a blue ocean market for thin-film modules. They might soon be found in tents, roll-able chargers and backpacks.

The above features make thin-film cells throughout their whole life cycle (25 years) generate 10% more power than that generated by crystalline silicon. Meanwhile, with domestic manufacturing, increased scale and improved technology of high-end thin-film equipment, thin-film products are more competitive than crystalline silicon ones by measuring cost per watt and cost per kWh.

Welcoming The Thin-Film Era

The transition to thin film is a process of upgrading solar technologies. In accordance with internationally accepted standards, solar cells can be classified into: first generation (polycrystalline silicon cells and monocrystalline silicon cells); second generation (low-cost a-Si thin-film cells and CIGS thin-film cells featuring high efficiency, low cost, high conversion rate, and large-scale industrialization); and third generation (organic solar cells and dye-sensitized solar cells and other emerging cells).

However, the biggest challenge is that many people do not know the difference between crystalline silicon and thin film. They simply make the mistake of thinking that the whole PV industry is an industry of crystalline silicon. A document issued by the China Banking Regulatory Commission said the government should not encourage but curb its lending to the development both of the PV industry and of the cement industry. Punishing these two industries indiscriminately is wrong. The crystalline silicon sub-sector does have many problems, but the thin-film sub-sector does not share them. The core of the third Industrial Revolution is a new energy revolution. The application of thin-film technologies can help distributed power generation in this revolution.

Thin film will become a mainstream PV technology worldwide. However, there are many people who still think crystalline silicon will continue to dominate in China. They claim that it's more difficult to develop and apply thin films. Naturally, funds flooded to the polycrystalline sector, which suffered from overcapacity as a result.

Why did makers of crystalline silicon in difficulty not turn to thin-film cells? The reasons are simple. First, they have no relevant technology. Second, the barrier to entry is very high—the investment per unit capacity is eight to ten times higher than that of polycrystalline. Third, the payback period in thin film is long, which prevents investors from making money in the short term.

Currently, thin film only represents 10% to 15% of the PV market, but we should not neglect its dramatic growth just because of its small market share. In the last five years, its market share has been increased from 1% to 10%, and even up to 15%, representing spectacular growth each year. In the next few years, thin-film cell makers will develop technologies, upgrade their equipment, and produce cells on a large scale after they expand their capacity. Some powerful enterprises have begun to master the core technology, and are capable of doing R&D in this sector. They are taking up an increasingly important position in the PV market.

Thin-film technology will become the mainstream, one which also adapts to the changes in PV application. In the future, BIPV applications will be more popular. Correspondingly, thin film and flexibility will gradually be widely adopted. BIPV applications will be the perfect marriage between solar modules and architecture. Even in urban areas with limited land, power generation will be practical. Distributed generation can ease the pressure on the grid during times of peak demand in cities, and it can improve the reliability and stability of the power system as a whole.

It is estimated that the potential installed capacity of China's existing and new PV buildings will reach 1 billion kilowatts by 2020, equivalent to the capacity of 368 Gezhouba Hydropower Stations or 45 Three Gorges Hydropower Stations. Promoting the BIPV industry on a massive scale will exert a positive and revolutionary impact on China's economic development. From raw materials to products, even VP to installation and application, the BIPV process is a typical case of industrialization. Especially, among the seven strategic emerging industries,[14] four industries are relevant to the thin-film cell technology for which China has independent property rights. These four sectors are energy-saving and environmental protection; advanced equipment manufacturing; new materials; and new energy. The thin-film industry is a technology-based sector that will come to occupy a real position in the economy. It will effectively promote industrial restructuring and upgrading, and lay a solid foundation for sustained economic growth.

Vigorously promoting the BIPV industry can optimize the existing power supply mix. Meanwhile, new energy as a necessity will replace fossil fuels in production and in daily life. In this way, the goal of stimulating domestic demand by consumption can be achieved. The imminent scale of the potential BIPV market is about 10 trillion yuan, 3 to 5 times bigger than the automobile market in China. It will generate indirect economic benefits of up to 30 trillion yuan, and pay fiscal taxes of 4 trillion yuan in total. According to conservative estimates, the BIPV industry will create stable jobs for at least ten to twenty million employees.

It is predicted that large-scale BIPV applications can help meet 30% of the annual power demand from society by 2020 (based on the standard of 2010). To be specific, 20% is from industrial production, 5% from households, and the other 5% from tertiary industries. In addition, a BIPV application of 1 billion kilowatts can potentially reduce carbon dioxide emissions by 1.3 billion tons per year, equivalent to 20% of the total emission in China per year (calculated based on an annual emission of 6 billion tons). Every year, the BIPV sector will create economic benefits of tens of trillions of yuan, ensuring the national economy can grow sustainably, and help reduce emissions and promote economic growth at the same time.

> Developing BIPV will benefit the country and the people. First, it can stimulate domestic demand. When households generate power for themselves, a PV investment leads to consumption. Second, it will help China meet its targets in emission reduction. The annual emission of carbon dioxide will decrease by one-fifth after adopting BIPV. Third, it will facilitate industrial restructuring and change the pattern of economic growth.

With current PV technologies, China is well prepared for integrating BIPV into the grids in the long run. The key is how to popularize BIPV on the market and change people's attitude towards energy usage. If the central government provides tax incentives to boost this industry in a compulsory manner, the cost of BIPV will be 0.5 yuan/kWh. Moreover, in sunlight-abundant areas, the cost will go down to less than 0.5 yuan/kWh. Grid parity can be achieved without any governmental financing. This will be a significant and meaningful energy revolution.

Based on pilot programs, the state can impose BIPV by introducing standards of energy conservation and emission reduction, especially targeting new projects. The government should encourage consumers to generate power for themselves through BIPV systems by asking the grids to provide two-way metering and net metering.

The PV industry across the world is facing an important strategic opportunity for transformation and upgrading. In the solar industry, China and other developed countries have stood at the same starting line. If the Chinese government clearly identifies thin-film technology, which is environmentally friendly, flexible and versatile, as its major direction to develop the solar industry, and gives support in terms of finance, technology and the wider market, China will be able to lead this third Industrial Revolution.

It is strategically significant for China's "steady growth, structural adjustment and industrial restructuring and upgrading" to realize China's strategic upgrading from crystalline silicon to thin film, to maintain its leading position in the field of thin-film solar energy, and to accelerate the development of the thin-film solar industry.

China's thin-film industry has demonstrated its position of global leadership. The government should mobilize financial and regulatory agencies to launch relevant financial incentives as soon as possible, especially for the manufacturing technologies for high-end thin-film equipment. Without that financial support, it's hard for the sophisticated leading technologies to be localized and China's leading technological edge over the competition will be useless, even though China owns independent intellectual property rights (IIPR).

At present, China has built a complete PV industrial chain and held the world's most advanced PV technologies, which makes China much more competitive than other countries. Meanwhile, China has ridden the global tide of PV development by seizing the strategic opportunity of technological upgrading towards thin film, and it has led the world in thin-film cell technologies through independent research and technology acquisitions. That is why we are so confident in China's leading position both in this new round of energy revolution, and in the third Industrial Revolution.

WHY CHINA? THE ADVANTAGEOUS POSITION OF CHINA'S PV INDUSTRY

In the near future, any and all applications of PV technology, from the PV power plants in Golmud, Qinghai Province, to the rooftop solar panels of households in Hainan Province, will become part of a larger Chinese PV market.

Boasting a territory of 9.6 million square kilometers, a population of nearly 1.4 billion, a bank balance of over 100 trillion yuan, and an annual economic growth of 9.8% for 30 consecutive years, China has become the world's second-largest economy.

China has a group of outstanding entrepreneurs who have grown up after the beginning of reform and the opening up of society. By making use of policy (a "visible hand") and the market (an "invisible hand"), China's PV technology has developed faster than Europe and the US, and so gained the competitive advantage in this sector.

China's PV industry has "ten advantages." We believe that China is grasping the opportunities brought by the new energy revolution and the third Industrial Revolution. China will tear off the label of being "a big country in PV manufacturing and a small country in PV applications" currently applied to it, and instead become a true PV leader in the world.

Breaking the Energy Bottleneck

China is already the second largest economy in the world, while its energy consumption (per unit of GDP) is more than twice the global average, and three to six times that of other developed countries, even higher than India. Energy has become a "bottleneck" and a "weakness," constraining China's sustainable economic growth.

Carrying out a new energy revolution and adopting a strategy focusing on new and alternative energies are the only solutions for breaking though the "energy bottleneck" and overcoming "energy weakness." The protagonist of this revolution is the solar PV industry.

Popularizing PV application is like killing many birds with one stone for China's dream of becoming a strong country. It can stimulate domestic demand by transforming the investment-driven economy to a consumption-driven one, promote industrial restructuring and the acceleration of the transformation of China's economic growth model, and help China achieve its goal of energy saving and reduction of emissions.

The Rising Giant's Energy Bottleneck

As the focus of global development has shifted to emerging markets, China has been leading these emerging markets. On May 16th, 2005, the Ninth Fortune Global Forum was held in Beijing. Eight years later, the forum was held in China again: on June 6th, 2013, Chengdu played host to the Twelfth Fortune Global Forum, with the theme being "China's New Future."

Why was the Fortune Global Forum held in China again so soon? Because this country has outperformed other emerging nations in a variety of aspects: a potential market with a population of nearly 1.4 billion, the rapid economic growth, the continuously improving infrastructure, the stable society, and the many opportunities for development. China has just become a middle-income country and is encountering many new opportunities. Meanwhile, China still adheres to its policy of opening up, and continuously deepens its reforms: this greatly appeals to foreign investors, both in terms of the specific policies, and their effect on the hardware and software markets.

It is necessary to have a look at China's excellent economic performance from 1952 to 2012. In a time of global economic growth, China made especially great achievements.

From 1952 to 2008, China's GDP grew 8.1% per year on average; its economic aggregate increased 77-fold; and China became the third largest economy in the world. The GDP was only 67.9 billion yuan in 1952, increasing to 364.5 billion yuan in 1978, and finally reaching 30.067 trillion yuan in 2008, with an average annual increase of 8.1%. China's economic aggregate accounted for 6.4% of the world total, behind only that of the US and Japan.[1]

Over the past five years, China's economic strength and comprehensive national power reached a new level. From 2007 to 2012, China kept growing 9.3% annually, as the world's second-largest economy. In 2012, its GDP exceeded 50 trillion yuan, and the per capita GDP reached 38,400 yuan. In 2010, China's national income per capita helped China gain a position among the "upper-middle income" economies.

Since 2013, the global economy has witnessed a lull, while China maintained its rapid growth relative to other emerging economies. After years of sustained and double-digit growth in GDP, China expects to achieve a year-on-year increase of 7% or 8% in the next decade, which is still higher than other economies. However, this indicates China has slowed down its economic growth by focusing on the quality of this growth. In fact, this also shows that the extensive growth model of the past was unsustainable.

In 2011, China surpassed Japan in GDP and became the world's second-largest economy. Meanwhile, China was also one of the world's largest energy consumers, especially of coal.

In 2010, China consumed 3.25 billion tons of standard coal. In 2011, China and India contributed 98% of the global net growth in coal consumption, and China became the biggest coal consumer in the world. In 2012, the energy consumed by China reached 3.62 billion tons of standard coal, equivalent to the total consumption of the rest of the world.

China is both a developing country and a manufacturing power, which justified China's high consumption of energy and low efficiency of energy use. At present, China's energy consumption per capita is equivalent to the world average, while the per capita GDP is only about half of the world average.

It will take a long time to reduce energy consumption per unit. However, with the expansion of the economy and the continuous improvement of people's quality of life, energy consumption will increase correspondingly, even though energy efficiency has been improved. It is estimated that China's annual energy consumption will reach 5.5 billion tons of standard coal in 2020; and 7.5 billion tons in 2030. In other words, the total energy consumption will keep growing gradually in the near future.

However, energy reserves in China cannot meet the country's increasing demand. It is estimated that according to the current exploitation rate, the proven oil reserves in China will be exhausted within less than 10 years, and those of natural gas and coal within 33 years.

In order to ease the tension between growing energy demand and its insufficient reserves, the State Council requires that total energy consumption be capped in China to below 4 billion tons of standard coal by 2015.

However, it is basically inevitable that China will accelerate its industrialization and urbanization processes in the coming years, and that China will maintain high energy demand throughout this time. To provide sufficient energy to meet this demand is a challenge. China will face more pressure on the sustainable development of its resources, ecology and environment.

The energy bottleneck has become the biggest constraint for China's economic sustainability. So what can we do?

> The only way to break the "energy bottleneck" and overcome "energy weakness" is a new energy revolution and the "strategy of substituting by new energy." Through this new energy revolution, China will be able to meet its energy demand by boosting the PV industry and changing the energy mix and system.

Why is energy supply the bottleneck and weak point in China's sustained growth? There are three inevitable problems. First, more energy will be consumed during economic growth. This connection between growth and energy is inevitable. Second, China's reserves of fossil fuels cannot meet its own demand. This reality is also inevitable. Third, this fossil energy consumption is bound to cause more and more serious environmental damage. This causality, too, is inevitable.

Here, we can see that these reasons make the bottleneck difficult to avert, or simply unavoidable. Why do the three causes exist? The consumption of fossil fuels can explain everything. If China stops or reduces its consumption of fossil fuels, these inevitable connections will no longer exist.

Therefore, the simplest countermeasure against the energy bottleneck would be the following: carrying out the new energy revolution, replacing fossil fuels with new energy, changing the energy mix, and making new energy dominate energy consumption. If inexhaustible solar energy can meet continuously increasing energy demand, the first connection will fall away. Solar energy is everywhere and it is green, so the other two causes will disappear also. Therefore, we can see that solar energy can easily solve these three problems.

In 2011, the coal used for power generation accounted for 53% of China's total coal consumption; that is, China burns 1.46 billion tons of coal for power generation in one year. If solar energy can replace at least 50% of this coal consumption, then coal consumption will be greatly reduced, and pollution will be largely abated.

In China's northern and coastal areas, annual sunshine duration reaches more than 2,000 hours. In particular, the sunshine duration in Hainan is over 2,400 hours per year. China is a truly sunshine-abundant country.

Boosting Transformation and Becoming Strong Through PV

It was snowing on December 16th, 2012, when the Economic Work Conference of the CPC Central Committee concluded in Beijing. American think tanks said that the Chinese government was undergoing a great reform: the transformation of its economic development mode. When the global economy was in a fundamental transition, China was extremely brave to make such a choice, one it made so that the reform would lay a solid foundation for China's growth over the next five decades. Only a smooth transition can help realize China's sustainable development.

China's economy is facing two transformations: the transformation of the "development mode" and the transformation of the "development model."

The "development mode" was formerly known as the "growth mode." In order to distinguish the concepts of "development" and "growth," and show that "growth" is not the same as "development," "growth mode" was changed to "development mode." At present, China's economy is developing in an extensive manner; boasting high input, high consumption, high emissions, low output and low efficiency.

In 2009, China's output of crude steel was 568 million tons, and its cement output was 1.65 billion tons, accounting for 43% and 52% of the world total respectively. And most of that was consumed by China. The primary form of energy consumed was the 3.1 billion tons of standard coal China used that year, accounting for 17.5% of the world's total. China's GDP over the same period was 34 trillion yuan, accounting for 8.7% of the world total. The ratio between the proportion of China's energy consumption of the world total, and that of output was 2:1, far behind the world average. All of the above shows that China is still in a stage of extensive development.

Since the outbreak of the financial crisis in 2008, the issue of the "development model" has become more and more prominent. China's current development can be described as depending on extensive expansion and foreign trade. The extensive expansion means increasing inputs to get more outputs, and the overdependence on foreign trade is a result of specific historic conditions. Since China's entry into the WTO in the early 21st century, it has broken through various trade barriers. China's products penetrated global markets with low costs and low prices, especially the markets of various developed countries. The shelves are full of goods labeled "Made in China." For nearly a decade, China's foreign trade has witnessed double-digit growth. China built up trade surpluses in many countries. Thus, since China became a manufacturing power, 70% of its economic growth has come to rely on foreign trade.

The global financial crisis in 2008 made this development model, which excessively relies on foreign trade, unsustainable. In order to overcome the crisis, developed countries forced China to appreciate its currency; they wanted to weaken the competitiveness of Chinese products. They frequently took protectionist measures to block the import of China's products. They also canceled a large number of orders for Chinese products. For example, in Dongguan,

Guangdong Province, thousands of enterprises, mainly relying on foreign orders, simply collapsed. This is just a microcosm of the affected regions in China, and the difference was only a matter of degree. The following is a fact: the higher the dependence on foreign trade there is, the greater the impact on business will be.

In this financial crisis, the Chinese asked themselves, "Why did the crisis spread to China, to the extent of having such a great impact on China's economic growth?" It comes down to China's development model relying to such a degree on extensive expansion and foreign trade, which has at least three drawbacks.

First, this model makes China consume more energy and resources, and causes more environmental pollution.

It was thanks to this model that China became a manufacturing power. The positive side of it is: compared with the past, China's economy *has* made progress. The negative side is: this development model is unsustainable. In human economic activities, manufacturing is the biggest consumer of energy and resources, and the biggest environmental polluter. If China maintains the model "Made in China, Consumed by the World," it will suffer severely from having insufficient energy and resources to handle such heavy consumption, as well as from the polluted environment caused by this kind of consumption.

Second, it is difficult for China to change the distressing situation of "buying expensive and selling cheap" on the international market. To meet the worldwide demand for products requires sufficient energy and resources, but China's own energy and resources are far from sufficient. It has to resort to importing a large amount in these areas. In order to ensure its large-scale production capacity, China had to import raw materials at high prices. For example, the price of oil per barrel on the international market has risen from several dollars to dozens of dollars, even to more than one hundred dollars. No matter how high the price is, China has no choice but to purchase more. Another example can be seen in the field of iron ore. Sometimes, iron ore prices increase by several tens of percentage points or even doubled, but China still feels incapable of bargaining. Those with resources have the final say. The export of China's products mainly relies on low prices. The more one makes, the more challengingly one prices its products. In addition, the bloody pricing war among domestic enterprises makes the price per unit produced lower still. And this trend has become irreversible. The model of "buying expensive raw material and selling cheap products" has minimized China's economic benefits. Over time, China will suffer losses.

Third, huge foreign exchange reserves cannot be used effectively. With the increase in foreign trade, China's foreign currency reserves accumulated rapidly, exceeding USD 3 trillion in 2011; reaching USD 3.31 trillion at the end of 2012; and totaling USD 3.5 trillion at the end of June, 2013, ranking first in the world.

This certainly marks China's economic rise, but it also concerns this country's sustainable growth, because that large an amount of foreign exchange reserves cannot be used efficiently. As the Western developed countries have not recognized China's market economy status, some countries even take up containment policies. When China wants to buy their high-end new technologies with foreign exchange, they refuse to sell. When China wants to purchase valuable corporate assets or energy and resources from them, they set up obstacles. A large amount of foreign exchange has to be reserved as US dollars in foreign banks, or be used to purchase US Treasury bonds. Since the "gold standard" was abandoned in the 20th century, the US has had more freedom to decide on the circulation and value of its currency, in line with its own needs. It can transfer its economic problems to other countries, and take other countries' wealth or even make it evaporate without being noticed. From this perspective, the larger the foreign exchange reserve China has, the less safe its economy will be.

Therefore, even without the financial crisis, the model that depends on extensive expansion and foreign trade is unsustainable in the long term. Thankfully, this crisis has made us realize this sooner rather than later. Our task is to change this distressing situation by taking certain initiatives.

In the context of the global economy, a successful economic transformation needs China to be well prepared in four key aspects. First, China must maintain a steady growth rate that must not be significantly reduced due to the transformation. Second, as the driving force behind foreign trade weakens, China must strengthen instead its domestic demand, including investment-led and consumption-driven demand. Third, China must realize intensive economic growth by restructuring its industry, upgrading its technology and improving their management. At the same time, it should improve the quality of both its macro economy and its micro economy. Fourth, China must reduce environmental pollution, and strengthen the construction of a truly ecological civilization.

Economic transformation requires many arduous, painstaking efforts. The key is to achieve a new economic growth point to meet the four requirements above. On July 15th, 2013, the economic data released by the National Bureau of Statistics showed that China's GDP was 24.8 trillion yuan in the first half year alone, with a year-on-year increase of 7.6%. The policies launched after 2013 indicated that the central government would tolerate a slightly lower growth rate, but required a better quality of growth. Tolerating the negative impact of economic slowdown to give more time and space for the transformation of the development mode may be "courage" in the eyes of US think tanks. Nevertheless, this tolerance has a limit. After all, the nation's development is our absolute priority. So a growth point that can realize both the desired economic growth rate and the required transformation of the development mode is highly anticipated.

I believe that the solar industry is one of the new growth drivers in China's transformation. How can the PV industry lead China's economic transformation?

First, this industry has a large market. It is estimated that in urban and rural China, there is a construction area of nearly 90 billion square meters that can be used for PV power generation. If the conversion efficiency of cells is 10%, and if 15% of eastern, southern and western walls, and 20% of rooftops are installed with cells, there will be 1000GW of installed capacity nationwide, approximately equivalent to the total installed capacity of thermal power, hydropower and nuclear power. It would also equal the capacity of 45 Three Gorges Dams. This will directly boost the market to the tune of 10 trillion yuan. BIPV (Building Integrated Photovoltaics) itself can bring 30 trillion yuan to the market, three to five times as much as the value contributed by China's automobile industry. In addition, there are the PV power markets for large-scale concentrated generation and energy-saving appliances (e.g., for mobile energy terminals), both of which have great development potential.

Second, this market can open up immediately. The feed-in tariff standard for large-scale centralized power generation has already been released. The distributed generation policy of "incorporating the spare electricity after home use into the grid" has been introduced. If the government can provide a certain degree of property tax relief, then new buildings under construction could be installed with power generation systems on a massive scale.

Finally, in addition to the first two reasons, the emerging solar industry can play a multi-faceted role in China's economic development.

Role one: the PV industry can stimulate domestic demand, shifting the PV market from an investment-led mode to a consumption-driven one. The PV market is a liberal one, where households can generate power for their own demands.

In May 2012, the Ministry of Housing and Urban-Rural Development issued the *12th Five-Year Special Plan for Building Energy Saving*, in which the applicable construction area of renewable energy buildings (mainly BIPV) was set at 2.5 billion square meters, 60 times larger than the 40 million square meters of the demonstration areas specified in the 11th Five-Year Plan's period. This indicates that BIPV will lead the PV industry and so usher in a boom.

Meanwhile, the Ministry of Finance made an announcement that the governmental subsidy for the new BIPV projects would be tentatively set at 9 yuan/watt, 3 yuan higher than it was in 2011.

Role two: China should actively promote the PV market, which will facilitate industrial restructuring and the transformation of the development mode. The most important words in the 2012 Central Economic Work Conference were "redefining China's strategic opportunity." It is no longer the traditional kind of opportunity that takes the form of simply becoming one part of the greater global labor market, expanding exports, and accelerating investment. Rather, it is a new opportunity which forces China to expand its domestic demand, improve its innovative capabilities, and promote the transformation of the economic development mode. It shows that China will never be taken for the "world's factory" again, and will never be shackled by the game's previous rule, the "theory of global labor division."

The term "world's factory" refers to the idea that the Western world makes use of China's resources and labor to serve themselves, while China has to consume a huge amount of its energy and resources. When China lacks energy and resources, it has to import them at a high price. For some major materials (such as crude oil, iron ore, etc.), import dependence has increased from 5% in 1990 to more than 50% in recent years. China pays huge amounts, but the return remains very low. "Excessive inputs and poor outputs" is the typical extensive development mode.

The crystalline silicon PV industry mirrors this. Due to the rapid development of overseas markets, especially the EU market, China's polycrystalline cell producers increased in number to more than 100 in a short time. In the 1980s, the annual production capacity of polycrystalline silicon in China was only 350 tons. But according to a survey of more than 50 polycrystalline silicon producers worldwide conducted by Solar&Energy, an energy market research institute, the total output of China's polycrystalline silicon will reach 165,850 tons in 2013, ranking first in the world. Silicon raw material is a kind of mineral resource. Despite increased investment, the output value of cells still keeps falling with the decreased price.

Developing BIPV will change this situation. With low energy consumption, zero pollution, flexibility and good weak-light performance, thin-film cells are gradually replacing crystalline silicon cells in BIPV applications. Although rooftops can be installed with both crystalline silicon cells and thin-film cells, facades and other building components are only suitable for these thin-film cells. Because the facades are not exposed to sunlight at right angles, only installing thin-film cells can ensure power generation under such weak light conditions. Building components such as curtain walls and glass windows can only adopt thin-film technology, because these have to be light and translucent. The current thin film technology is the second generation of such technology, and it belongs to the sphere of high-end technology. Replacing crystalline silicon with thin film is both a technological upgrade and a shift in the development mode.

Role three: developing the PV application market will help achieve the goal of energy saving and emission reduction. Before the Copenhagen Conference, China had made specific commitments on emission reduction: by 2020, carbon dioxide emissions per unit GDP must decrease by 40% to 45% compared with those of 2005. It requires tremendous effort to cut such an amount of carbon dioxide emission in such a short time. It is estimated that if China popularizes BIPV nationwide, then carbon dioxide emissions will be reduced by 18%, around 1.3 billion tons, annually.

We must recognize that it is economic transformation and energy constraints that urge China to choose PV. In the new energy revolution, China must be able to make a difference.

China's Unique Growth Pattern

China's solar PV industry has "ten advantages," among which strategic advantage and the opportunity advantage are the most important. Taking thin film and flexibility as its strategic direction, and boasting of world-leading technologies and enterprises, China has grasped the opportunities offered by the new energy revolution and the third Industrial Revolution.

The "China model"—with its unique and relatively efficient system and positive policies—becomes the third important advantage. Since 2012, the Chinese government has made rapid moves to introduce favorable PV policies, enabling the troubled Chinese PV industry to find its way out of its rut at last.

Strategy, Opportunity and the "China Model"

I am a veteran who has been fighting in the new energy war for more than two decades, but I am also a CEO who is increasingly confident about the promising solar PV industry.

Nevertheless, I am more like an orator sometimes: I introduce the development strategy of this industry to government officials and experts; I convince my counterparts to develop thin film as I do; and I popularize our corporate mission and the significance of engaging in the PV industry among my employees.

On several different occasions, I have publicized the "ten advantages" of developing the solar PV industry in China—namely strategic advantage, opportunity advantage, institutional advantage, policy advantage, market advantage, industrial advantage, technological advantage, talent advantage, capital advantage, and cost advantage—all detailed below.

China has ten advantages in developing its PV industry:

1. Strategic advantage: thin film and flexibility are the direction and trend of PV development; and China is a leading player in terms of technologies and businesses.
2. Opportunity advantage: the era of thin-film replacement on a massive scale is coming; China is well prepared in terms of technology and other key elements; and China is capable of turning environmental pressures into industrial drive.
3. Institutional advantage: China has a unique and relatively efficient system, and mechanisms.
4. Policy advantage: China encourages and supports the development of the real economy, including the solar industry.
5. Market advantage: China's PV market has opened up, and it will become the largest in the world.
6. Industrial advantage: China's manufacturing is powerful.
7. Technological advantage: Hanergy with its global technological integration has helped China's thin-film PV industry achieve a leapfrogging development model, and thus a leading position in the world.
8. Talent advantage: besides PV technicians, China boasts of a group of outstanding businessmen who embody entrepreneurship.
9. Financial advantage: China has abundant private capital; household savings are above 43 trillion yuan; and Renminbi deposits exceed 100 trillion yuan.
10. Cost advantage: the global integration of technologies is successfully combined with products "Made in China."

China has grasped the historic opportunities afforded by the third Industrial Revolution and the new energy revolution, in which new energy is becoming the substitute for rapidly depleting fossil fuels. The upgrading, transformation and sustainable development of China's economy have positioned solar energy instead of other energies as the main energy resource. This is the "opportunity advantage."

As for the strategic advantage, I have mentioned it before to some extent; the European Commission is grappling with the economic crisis; the US is developing shale gas; and both are losing their strategic direction to some degree. Japan and South Korea have their own limitations. Only China, the PV manufacturing powerhouse, is qualified to be a strategic PV leader. Through the integration and upgrading of the polycrystalline silicon pattern, China has understood that thin film and flexibility are the right direction and the general trend of global PV development. In addition, China owns leading technologies and competitive companies in this regard.

For the market-and-industry, technology-and-talent, and capital-and-cost advantages, I am going to provide details in the succeeding parts of this chapter. Here I will focus on the institutional and policy advantages.

The institutional advantage refers to how China's institutions and mechanisms are unique and relatively efficient. The policy advantage indicates that the Chinese government encourages the real economy by releasing favorable and effective policies. The underlying reason is to promote rapid economic growth in the context of the China model.

Since the reform and subsequent opening up, China has achieved rapid economic growth. All its tremendous achievements can be attributed to an unprecedented growth model, the China model, which successfully combines traditional market mechanisms with macro-level control.

There are numerous varieties of Western economic theory, which can be classified into two main schools: one is against government intervention, and believes in completely relying on the market to solve problems; the other, notably Keynesianism, admits the market's role in problem solving, but believes that the government can and must intervene appropriately. Putting the academic debate aside, we find in reality that economic intervention by the government is essential. The question is how to intervene. The fact that the US tided over the 1929 economic crisis is generally recognized as a good example of President Roosevelt using Keynesian economic theory to governmentally intervene during an economic crisis. Since the outbreak of the financial crisis in 2008, President Obama has been forcefully intervening where he deems necessary. It is true that in this new energy revolution also the government needs to strengthen its guiding role.

Being inconsistent with Western free market economics, the China model is questioned by some people, because the Chinese government is often regarded as the only authority in China. However, "authority" can be an advantage as long as it plays an active role. Besides, the China model is market-oriented, which means that the market decides resource allocation, and then the government exerts macro-level control.

Since the 1990s, by relying on macroeconomic regulation and control, China has weathered economic overheating in 1993, the insufficient domestic demand and Asian financial crisis of 1997, as well as the global financial crisis in 2008. Over those years, China has maintained its rapid and sustained development.

In 2012, the key to the quick recovery of China's ailing PV industry was also that the government played a good guiding role again.

The features of the new energy revolution determine that its evolution is different from general economic activities. The biggest difference is that the government must strengthen its leading role while following the laws of the market economy. If there is no helpful guidance from the state and government, and if China lets the market take its course without any checks or balances, it will be hard to make any kind of a breakthrough. No country in the world is an exception.

China's economic system determines that it's easy for the China model to converge with the new energy revolution.

When the "group crisis" of China's PV industry erupted in 2012, many people focused on the interaction between the government's role and the market mechanism. Some believed that overcapacity in the PV industry was related to some local governments' "offside," and that it was the deep involvement and excessive support from the local governments that caused PV to overheat. Others said that this was due to the government's lack of such influence. For the whole industry, the government lacked appropriate planning and guidance. Given the potential risks of the model relying on extensive expansion and foreign trade, the government failed to exert control or to provide guidance in a timely manner.

These views seem reasonable, but do not truly get to the heart of the matter. They fail to fully understand the unique and relatively efficient features of China's institutions.

In China's energy sector, whether it is primary energy or secondary energy sources, whether it is sole proprietorship, joint-stock companies or listed companies, most firms are controlled by state-owned capital interests. Some think that such ownership is potentially detrimental to the energy revolution. On the contrary, the essence of state ownership is that all the people are the ultimate owners and shareholders. Although the state-owned companies should strive for more profit for themselves, their fundamental purpose should be to lead the sustained and healthy development of the national economy, and to safeguard both the national economy and people's livelihood. Such entities in the energy economy, as a part of its basic make-up, should not block the new energy revolution just for the sake of their own vested interests, because the revolution will benefit the long-term development of the entire national economy. Those entities should become active participants and main forces in the new energy revolution.

From a practical point of view, the ultimate decision-maker for state-owned enterprises is the government. They must comply with the policies and directives of the government. For example, on October 26, 2012, the State Grid, which monopolizes the transmission, distribution, and sale of electricity, issued the *Opinions about Providing Good Grid Connection Services for Distributed Photovoltaic Power Generation*. The quick and favorable response of the State Grid proves my point.

The Chinese government is able to "concentrate power to do something big." This provides a good institutional environment for China's new energy revolution. That is China's institutional advantage.

Now, let us discuss the policy advantage. Policy is an important engine for the PV revolution. In November 2011, Hanergy invited experts from various fields to discuss how to promote the new energy revolution in China. At the end of the meeting, a senior expert concluded, "Now everything is prepared except one thing. The 'thing' is policy."

Starting a project needs more power than just running it. The new energy revolution is no exception. When summarizing the development of the European PV industry, I mentioned that their PV industry has successfully gone through a subsidized "infancy" and entered the growth stage of market operation. China's PV industry is in such transition, and really needs policy support.

At present, China's PV industry is like a car that is climbing up a hill. The slope is so steep that the car cannot make it even if it goes full steam ahead. If the passengers help push the rear, the car will get on the flat road, and the passengers can then enjoy being carried forward. If the government pushes the climbing PV "car" with its policy decisions, then it will enjoy the economic and social benefits brought by a developed PV industry.

The so-called "policy in place" is not to simply develop a specific policy, but to solve a series of problems, and to establish an inter-consistent policy system. For example, how should finance, tax and price play a leverage role to ensure the healthy development of this industry? How should we determine policy priorities so as to ensure good performance in production, construction and application? How should China deepen the reform of the power system and adjust the relationship between power generation and grid connection, so as to realize a seamless docking between new energy enterprises and traditional energy giants?

Fortunately, China's power system has taken a big step in promoting large-scale grid connection of new energy power generation (such as solar energy), especially in the construction of distributed power plants. Many insiders have begun to prioritize the development of a "smart grid." In many pilot areas, PV power plants have started to try cooperating with the State Grid. These things show that the policy advantage is an integral part of China's PV development.

"The development assigns a topic, and the reform writes the article." Policy adjustment can promote development. In a sense, policy adjustment is the field that best reflects whether a government is strong, or merely charming but replaceable in its role of guiding the new energy revolution.

The national policy of encouraging the development of the real economy is also a strong pillar of support for China's PV industry. The year 2012 was not only a troubled year for this industry, but also a poor year for China's real economy at large. Medium- and large-sized industrial enterprises maintained negative growth in profits until September, when the national policy of stabilizing growth took effect and those enterprises started to turn a profit at last. It is clear that the government introducing various incentives for the real economy can have significant effect.

China's macroeconomic policy in 2013 was to highlight restructuring, to safeguard and improve people's livelihoods, and to focus on expanding domestic demand, creating more opportunities and a more relaxed environment for the real economy, effectively dealing with the important issues involving people's livelihoods. The CPC Politburo meeting proposed: "We must prioritize the quality and efficiency of our economic growth." This injunction puts more emphasis on the quality of growth rather than its speed, while the quality of growth largely depends on the sound development of the real economy.

In June 2013, when chairing a State Council executive meeting, Premier Li Keqiang also pointed out that financial institutions would serve the real economy better by optimizing the allocation of resources and making good use of both existing and additional monetary and financial resources. He also said that financial institutions should strengthen their credit support (especially for advanced manufacturing, strategic emerging industries, labor-intensive industries and service industries as well as the upgrading of traditional industries).

Enterprises are the dominant force for developing the real economy. Making a profit is the goal and responsibility of entrepreneurs. The first impetus to attract more investment in the real economy is to enable enterprises to profit from the real economy and gain higher ROI than from the virtual economy. The government should appropriately channel and allocate market resources through structural tax and other policy tools. On the one hand, the government can strengthen support for real industry, maintain industry's enthusiasm, and create a favorable development environment for the real economy; on the other hand, it can create a new security mechanism for production factors, prevent raw materials, resources, labor and other elements from overlapping, and thus increase profit margins.

All these favorable conditions should drive China to maintain its leadership in the PV industry.

The "Visible Hand" Has Made a Move

As we know, the economy has two hands. The market is the "invisible hand," and the government is the "visible hand." An economy that only uses the "visible hand" is a planned economy. An economy that rejects the "visible hand" belongs to the school of economic libertarianism.

> The correct relationship between the two hands should take this form: the invisible hand is the foundation, and the visible hand is the supplement. When sticking to the decisive role of the invisible hand, we must make good use of the visible hand to correct for the blind spots and weaknesses of the invisible hand. That is to say, improve the performance of the invisible hand with the visible hand acting in line with reality. That's the best scenario.

The market economy mainly depends on the invisible hand to allocate resources. However, it is wrong to say that the visible hand is useless or cannot be used altogether. The two hands have different roles and strengths in dealing with various economic problems at different stages of economic development. In economic crises, it is difficult to recover by entirely relying on the market mechanism itself, and it is better to depend more on government intervention to resolve economic problems. When an industry is in its infancy, the best development opportunity may be missed if it relies totally on the market mechanism. It needs the "visible hand" and policy support from the government to live up to its potential and so play a bigger role.

China's PV industry is a new emerging industry that could affect the economic development of the whole country, and its smooth development not only needs the help of the powerful invisible hand, but also asks for the visible hand to provide strong policy guidance, industrial protection and even diplomatic support. It is encouraging that the visible hand has made a move and come out to strongly support China's PV industry.

For China's PV industry, the year 2012 was a cold year, but a turning point as well. The state issued a series of regulations and supportive policies.

On May 23, 2012, the State Council executive meeting proposed to "support the use of self-contained solar energy products in public spaces and households." On September 12th, the state issued the *12th Five-Year Plan for Solar Energy Development*, expanding the installed capacity of PV power generation from 21GW up to the range of 30GW to 40GW. As one of the industry's most important financial backers, China Development Bank also proposed further strengthening financial and credit support for healthy development of the PV industry. On September 14th, the National Energy Administration issued the *Notice on Applying for Large-scale Usage Demonstration Area of Distributed Solar PV Power Generation*. On October 26th, the State Grid issued the *Opinions on Providing Good Grid Connection Services for Distributed Photovoltaic Power Generation*.

On November 9th, the Ministry of Finance, the Ministry of Science and Technology, the Ministry of Housing and Urban–Rural Development, and the National Energy Administration jointly issued a notice about launching a second round of Golden Sun and Building Integrated PV demonstration projects in 2012, so as to support the technological integration and application demonstration of PV-generation-oriented micro-grids. The government would provide financial assistance in accordance with investment. The notice also encouraged the installation of PV power generation systems in schools, hospitals, communities, public buildings and the like. Various regions were told to choose whether or not to apply for projects that qualified for the investment grants, and to try out electricity subsidies.

On December 19th, former Premier Wen Jiabao held a State Council executive meeting to discuss policy measures for the healthy development of the PV industry. The meeting identified five major measures: accelerating industrial restructuring and technological progress; regulating the industrial development order; actively developing the nation's domestic PV market; developing favorable policies; and giving full play to the market mechanism, reducing government intervention as well as prohibiting local protectionism. The conference was to become a milestone in the development of China's PV industry. It also indicates that China will never be satisfied with merely helping to develop the PV industry, but has put this industry in a position of paramount importance within China's entire energy strategy, and its national economic strategy.

In 2013, the national policy continued to strongly support the development of the PV industry.

In February 2013, the State Grid issued the *Opinions on Providing Good Grid Connection Services for Distributed Power Supply*. This document extended the service for distributed PV power generation to all types of distributed power supply, launched specific measures in this regard, including strengthening the construction of supporting grids, optimizing grid-connected processes, simplifying the procedures for grid connection as well as improving service efficiency, and also provided green channels and favorable conditions for grid connection.

On June 14th, one important item on the agenda of the State Council executive meeting was to discuss the status of the PV industry and make a breakthrough. On July 15th, the State Council issued *Several Opinions on Promoting the Healthy Development of the Photovoltaic Industry*, which became a programmatic document for the development of the PV industry. This document once again aroused the enthusiasm and hope of the whole Chinese PV industry. Within six months, the State Council executive meeting had discussed the same industrial issue twice, which fully demonstrated the importance of the PV industry.

On July 24th, in order to implement *Several Opinions on Promoting the Healthy Development of the Photovoltaic Industry* as issued by the State Council, the Ministry of Finance issued the *Notice on Subsidizing Distributed Solar PV Power Generation according to the Quantity of Electricity and Related Problems*, identifying specific measures in this regard. In August, the National Energy Administration released the first list of 18 distributed PV demonstration areas covering seven provinces and five cities. On August 22nd, the National Energy Administration and China Development Bank jointly issued the *Opinions on Supporting the Financial Services for Distributed Solar PV Power Generation*, to financially support the PV industry.

On September 30th, the National Development and Reform Commission issued an extremely important document, the *Notice on Improving the Healthy Development of Solar PV Industry by Utilizing the Price Leverage*. It said that according to the different levels of solar radiation in each region, the feed-in tariffs for new ground-mounted PV plants were classified into bands of 0.9 yuan/kWh, 0.95 yuan/kWh and 1 yuan/kWh. In addition, all the electricity generated by distributed solar PV power plants is to be subsidized at 0.42 yuan/kWh.

These policies pointed the way for the development of China's PV industry. In the gradual improvement of this industry, the visible hand continues to facilitate the development of China's new energy industry, especially the PV industry.

Advantage Element 1: Huge Market with Generous Funding

The capacity of the Chinese PV market outruns imagination. Reports say that China consumed 4.69 trillion kWh of electricity in 2011. If solar power plants can generate electricity for 1,300 hours each year, a total installed capacity of 3,608 GW could meet consumption demand at 2011 levels. However, China's installed capacity of solar energy was only 8.2GW in 2012.

The potential direct scale of China's BIPV market is worth about 10 trillion yuan, which is three to five times bigger than China's automobile industry, and it can contribute 30 trillion yuan to national economic growth.

Over the past 30 years, China has maintained an average growth rate of 9.8%. In the next 10 to 20 years, China is expected to keep growing at 7% to 8% per year. By the end of June 2013, the total deposited yuan balance exceeded 100 trillion yuan, and foreign currency reserves reached USD3.5 trillion.

The Huge Market Worth 30 Trillion Yuan

In the vast Gobi area in Qinghai Province, the PV power plants along China's National Highway 109 look quietly spectacular. Against the background of snow-capped mountains under the winter sunshine, rows of dark blue PV panels are dazzling.

Qinghai Province was chosen by China's State Council as the experimental base for PV power plants. Construction of PV power plants is well underway in the plateau. On the Golmud plateau, you can see thousands of workers constructing plants around the clock.

Qinghai's leadership is quite visionary. The first concession in the PV power generation project was in Gansu Province rather than in Qinghai. However, in May 2011, a well-known story circulated in the industry, saying that the Qinghai provincial government was unofficially promised that PV power plants which were connected with the state grid prior to September 30, 2011 would be subsidized to the tune of 1.15 yuan/kWh. This was known as "Project 930" among industry insiders.

At that time, the cost of PV power generation had been reduced to less than 1 yuan. That means the profit margin for generating 1kWh of electricity is 0.15 yuan, and the return is as high as 13%. Earlier, 26 companies had registered their local branches and conducted market investigation and research. Although many companies had signed letters of intent, few companies had made final deals and most of them still took a wait-and-see attitude. By the end of August, the total installed capacity was less than 0.04GW. After all, the on-grid price of 1.15 yuan was just an unofficial quote from the Qinghai provincial government.

Two months later, the National Development and Reform Commission (NDRC) announced the on-grid tariff, which was precisely 1.15 yuan. It meant that "Project 930" and the on-grid prices of Qinghai Province were approved by the NDRC.

When Hanergy was invited to invest in Qinghai in July 2011, there were more than 30 PV companies there. In 2011, the incremental grid-connected PV capacity in Qinghai was 1.003 GW, accounting for 50% of the total installed capacity of the country, and it was almost twice as high as the 0.53GW of newly installed capacity nationwide in 2010.

This is a story about a PV pioneer in western China; the kind of story that is now spreading from Ningxia to Gansu, from Inner Mongolia to Xinjiang and Tibet.

Solar energy development in western China is a microcosm of the huge Chinese solar PV market.

Through an economic lens, China's scale keeps expanding, and the demand for energy will definitely continue to grow. On March 19, 2013, China's economist Cheng Siwei said at the Fifth International Petroleum Summit that China was facing multiple challenges in energy supply, price volatility, and environmental problems. The global economy grows at 4% and energy consumption increases by 2.8% annually. Over the next decade, if China keeps an average 7% growth rate, it will likewise consume more energy at an annual increase of about 5%.

From the perspective of energy mix and distribution, China's energy shortage cannot be fundamentally addressed by solely relying on energy imports and such major projects as the West–East power transmission project and the West–East natural gas transmission project. China must substantially replace fossil fuels with solar energy. Solar PV energy will gradually replace fossil fuels and plays an indispensable role in the new energy revolution. In 2011, China's total power consumption was 4.69 trillion kWh. If the solar power plants generated power for 1,300 hours each year, China's consumption would require a total installed capacity of 3,608 GW.

I believe clean energy will account for 50% of the total primary energy mix internationally by 2035. China had an installed solar capacity of only 8.2GW in 2012. Clearly, it is a huge market with promising potential.

The general guideline for China's PV power plants emphasizes a gradual transition from centralized PV generation to a distributed generation model. In the future, distributed power generation will be the mainstream, and large-scale centralized power generation will only be a supplement. However, at the beginning of the third Industrial Revolution, large-scale centralized power generation is still necessary. PV power generation, as the main force and pioneer of the new energy revolution, will be indispensable in the replacement of fossil fuels. In 2010, China's investment in clean energy reached USD 54.4 billion, making it the largest in the world followed by Germany's USD 41.2 billion and the US's USD 34 billion. In the same year, the newly installed PV capacity in China realized a year-on-year increase of 66%. According to Goldman Sachs and some other research institutions, Europe has represented 70% of the global PV market in the last few years, but China is expected to be the world's largest solar market soon.

Another major marketable application for such technology is mobile energy terminals.

Imagine this: in his warm tent somewhere on Mount Everest in 2030, Mike is browsing on the Internet to check global ranking in solar radiation—the sunlight that powers PV technology. He found that Tibet, China, ranks second, and the Sahara Desert is first. The laptop that Mike uses is thinner and lighter than Apple's current tablets, and it needs no external power supply. His laptop can independently and continuously power itself by utilizing solar radiation. With the most advanced flexible thin-film solar technology in the world, the thin film can be easily bent to any shape he desires. The tent is also designed to function as a solar power plant that provides power for his electric blanket. When Mike gets out of his tent, he won't feel cold in snowy or windy weather, for his clothes are functioning as small solar power plants as well.

There will be a wide range of solar products, such as solar stereo, solar tents, solar watches, solar mobile phones, solar computers, solar fax machines, solar printers, solar lamps, solar cups, solar microwave ovens, solar ovens, solar hats, solar doorbells, solar monitors—and these are just the tip of the iceberg. For instance, a solar table would look like an ordinary table with a flat glass top, but that top is actually a piece of thin-film solar panel. Under the top, there are several plugs in the corners for charging mobile phones, computers, and other household appliances.

Mobile energy terminals are expected to push solar PV energy into becoming the ultimate substitute. Who will be the main consumers during this replacement process? They are the generations born in the 1990s and the 2000s. This process will be a process that pushes the third Industrial Revolution from the burgeoning stage into a golden era. This process will see us witnessing the younger generations taking a place of dominance in the world. Their values will determine the historical processes of the third Industrial Revolution.

"High technology, large scale, wide application" are the prerequisites for the development of the PV industry. "Wide application," or a huge market, is the most important one because high technology relies on large-scale manufacturing, which in turn depends on a big market to generate profits.

> Without an enormous market, it is difficult to create and maintain high technology and large-scale manufacturing. China has exactly that—an absolutely huge market!

Investing in The Photovoltaic Industry

Over 30 years since China's reform and opening up, China maintained an average annual growth of 9.8%. In the next one or two decades, China could maintain a growth rate of 7% to 8%. Some institutions have predicted that China will surpass the US in terms of GDP, becoming the world's largest economy by 2025 or 2030.

The underlying reason for this prediction is simple: being in the mid-to-late stages of industrialization, and in the middle of rapid urbanization, China's economy is still in the prime of its adolescence. China is a big country with vast territory and a large population of nearly 1.4 billion. It has the most abundant labor resources on the planet, and is highly competitive in global manufacturing. More importantly, it has the most potential and the most promising domestic market. Being a PV latecomer is a kind of advantage, and by settling the matter in one go, China can directly introduce the most advanced technologies in the world.

When the 2008 financial crisis broke out, some people were worried that there would be turmoil in China. However, China firmly disagreed with that hypothesis, for the fundamentals of China's economy remained unchanged. Over the past few years, the repercussion of the financial crisis could still be felt. Subsequent to the European debt crisis, the world economy has yet to return to its past vigor. Within this context, the Chinese economy still maintains a growth rate of 7% to 8%. The 18th CPC National Congress set two goals for the period from 2010 to 2020: doubling the GDP and the per capita income. To achieve this, China's economy has to maintain an average annual growth rate of above 7.6%. These are China's economic fundamentals.

China has an investment-driven economy. The destination of investment relates to current GDP growth. Furthermore, it influences the long-term, healthy and sustainable development of China's economy.

Since the outbreak of the international financial crisis, China has implemented a proactive fiscal policy by injecting 4 trillion yuan into infrastructure construction, including railways, roads and airports. As a result, the country's GDP increased by over 8% in 2009. However, this momentum was maintained for just one year before slumping into a continuing decline. During April and May 2012, China increased its investment again. Some regions even called for "working hard for one hundred days, investing heavily and constructing massively to achieve greater prosperity," but the results were not satisfactory. Although the fourth quarter of 2012 saw an increase of 0.9%, the first quarter of 2013 witnessed a decrease once again. Both the law of diminishing returns on investment in economics and the reality of the situation show that such short-term bailouts are losing their efficacy.

Real estate markets can attract huge investment and absorb vast amounts of capital. Although adjustment and control policies for the real estate market have been implemented for many years, no obvious effect is seen at all. It seems that the more policies are rolled out, the heavier the investment is. In May 2013, among the 70 cities in China, there was only one city that saw falling housing prices. The enthusiasm for real estate investment has maintained its momentum. According to the National Bureau of Statistics, over the first five months of 2013, the investment in real estate development was 2.68 trillion yuan, representing a year-on-year increase of 20.6% and only a slight symbolic drop of an increment of 0.5% if compared with the first four months of 2013. In addition, while that 2.68 trillion yuan was channeled into real estate development, there was more capital also injected into other investment activities and speculation that went to the second-hand (stock) housing market. High housing prices have become a hot societal issue. To curb soaring housing prices, the Chinese government has launched a series of policies to continuously and firmly contain investment demand, and dampen real estate speculation.

The automobile industry, especially in the premium car sector, has also been a hot sector for investment in the past few years. Since 2011, China has become the world's largest seller of vehicles. Vehicle ownership has reached 100 million total vehicles. Considering the saturated car market in first-tier cities and the environmental pollution caused by vehicles, some big cities have placed limits on car ownership. The relevant overcapacity in the vehicle industry will greatly limit future investment there.

Which industries should absorb more investment? How should capital be allocated more efficiently? As previously mentioned, the new economic growth point in China's transition period is definitely the emerging solar industry. The financial industry should serve the real economy, and focus more on supporting the PV industry.

Since the launching of anti-dumping and anti-subsidies duty investigations by the US and the European Commission, the most acute problem faced by Chinese PV companies is the financial issue. Because of inadequate cash flow, companies are confronted with huge financial pressure. It has been reported that the top ten producers of polycrystalline silicon in China had liabilities of more than 110 billion yuan. Financial institutions foresaw potential risks and cancelled loans to those producers. On March 18, 2013, the creditors of Suntech jointly submitted an application to the Wuxi Intermediate People's Court for the company's bankruptcy and reorganization. Meanwhile, LDK's operating conditions at that time still had not gotten any better.

However, against the backdrop of the shrinking global market, the total clean energy investment in China reached USD 67.7 billion in 2012, making China the world's renewable energy leader. The rapid development of the solar energy industry made relevant investment hit a record high with a year-on-year increase of 20%. Total investment surpassed that of the second player, the US (USD 44.2 billion), by more than 50%. With the government introducing a new series of measures, the growth rate of PV installed capacity could reach three digits in 2013, and this is just the beginning. According to government predictions, PV installed capacity will reach 40GW by 2015. If calculated as 8 yuan/watt, the total investment will reach 320 billion yuan.

In essence, the reason for the funding crisis caused by the anti-dumping and anti-subsidies duty investigations is that financial institutions are less than confident about the "having raw material imported and products exported" model which is adopted by some PV businesses. Feeling "something bad" is about to happen, institutions tend to pull back on lending.

The PV industry is both technology-intensive and capital-intensive. The investment at the outset is heavy, and the payback period is prolonged. Capital investment covers the entire industry chain—from plant construction and equipment-purchasing to raw material refining and purchasing, to the production and manufacturing of cells and modules, and finally to the construction and operation of power plants. The recovery of funds actually has to wait until the power plants start to generate electricity and collect bills. Any PV enterprises whose operations cover the whole PV industry chain have to be patient with the long period of cash flow. Even enterprises that are partially involved in the chain can also be affected and restricted by this issue.

In fact, China has adequate capital. Statistics released by the People's Bank of China on July 12 of 2013 showed that China's RMB deposits had hit 100.91 trillion yuan as of the end of June, exceeding 100 trillion yuan for the first time. Foreign exchange reserves had reached USD 3.5 trillion.

In recent years, Renminbi deposits have maintained rapid growth, surpassing 50 trillion yuan in May 2008, 80 trillion yuan by the end of 2010, and 90 trillion yuan in November 2012. In June 2013, RMB deposits increased by 1.60 trillion yuan. For the first half of 2013, total RMB deposits soared by 9.09 trillion yuan.

Besides healthy amounts of funding, China's domestic liquidity is adequate. In June 2013, China issued 860.5 billion yuan of new loans. New loans amount to 5.08 trillion yuan in the first half year of 2013. Within the same period, aggregate financing to the real economy was 10.15 trillion yuan, up by 2.38 trillion yuan compared with the year before.

By the end of June 2013, China's broad money amounted to 105.45 trillion yuan, which was a year-on-year increase of 14.0%, 1.8% lower than that of May, and 0.2% higher than that at the end of 2012. The year-on-year growth rate of narrow money at 9.1% was 2.2% lower than that of May, and 2.6% higher than that in June 2013, the end of 2012.

Advantage Element 2: Industry, Cost and Technology

China's reform and opening up has brought unparalleled industrial and cost advantages to China's PV industry, which make up the "reform bonuses" we are currently enjoying.

China has surpassed the US and the Europe in terms of its core competencies, which are represented by such technical indicators as conversion efficiency, cell production, and future installed capacity. China leading the world's solar industry is a reality that is already on the horizon.

Reform Bonuses: Industrial and Cost Advantages

Internationally speaking, the low cost of China's PV industry is a strong competitive edge. This cost advantage is due to China's industrial environment—China is a huge manufacturer.

Around July 2013, a graph indicating a reverse of the global manufacturing pattern, together with two news stories, were widely spread on the Internet. First, HSBC Purchasing Managers' Index™ (PMI™)[2] posted an eleven-month low of 47.7 in July, down from 48.2 in June, signaling a deterioration in business conditions for the third consecutive month. Second, according to the flash Market Euro zone PMI released on July 24, the index bounced to an 18-month high of 50.1 from 48.7 of June.

As shown in Figure 4–1, the Chinese PMI curve fell below the Eurozone PMI curve. This aroused market concerns and brought the advantages of China's manufacturing up for discussion.

Over the past decade, globalization experienced in-depth development, and the global industrial structure went through a new round of massive and significant adjustments. In the developed world, new technologies that center on information have been adopted widely and led the industrial structure to become more technology-intensive, knowledge-intensive, and service-intensive. Institutional innovation, which is characterized by deregulation, enhanced competition, and free moving factors, provides more room for industrial development and global transfer of such. The idea of sustainable development, which is based on energy conservation and environmental protection, has been widely accepted and posited higher requirements for industrial adjustment to realize corporate social responsibility objectives. Industrial restructuring and upgrading has

Figure 4–1 The PMIs of major economies in the world.

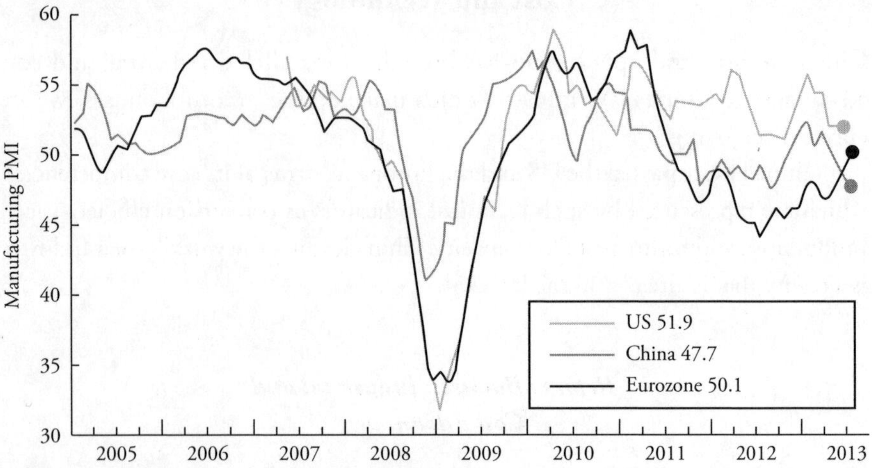

Source: HSBC China Manufacturing PMI Report, published by HSBC in July of 2013

greatly contributed to the further optimization of this growth pattern. In developed countries, technological advantages and industrial competitiveness have been enhanced to reduce energy consumption and pollution emissions, and technology and added value have been prioritized to promote economic growth. For developing countries, the global industrial transfer is a perfect opportunity and a major challenge for their transformation from extensive economies to intensive ones.

In its more than 30 years of reform and opening up, China has seized the opportunity in international industrial transfer, actively carried out institutional reform and mechanism innovation, achieved magnificent growth when taking on such transfers, and became the world's factory by replacing Britain, the US and Japan.

China has an abundant and high quality labor force. It has established a complete industrial chain and strong industrial bases, and nurtured a number of innovative leading companies whose creative capability, scale and brands are among the best in the world. China will prioritize development of the seven strategic emerging industries, such as new energy, biotechnology and the new generation of information technology. Thus, China is hopeful of making breakthroughs in major technologies, commercializing its major scientific and technological achievements, creating innovation clusters that possess international competitiveness, and gaining the commanding heights in emerging industries through leading technologies.

Taking PV cell production for example, we can compare China with the US, Europe and Japan.

Cell production capacity represents an economy's PV strength in mid-stream solar manufacturing. In 2005, the production capacity of Japan was 0.762 GW, accounting for 46% of total global output, and ranking first in the world. The market shares of the European Commission (second) and the US (third) were 28% and 9%, respectively. However, by 2010, China's solar PV production accounted for 47.8% in the world, replacing Japan as number one. The EU and Japan together took nearly 30% of the market share, ranking second in global PV production. The US, with a market share of 5.74%, took third position.

Along with this, the cost of Chinese cells and modules has been reduced so rapidly that it has become an advantage. In 2006, the cost of PV cells and modules hit USD 5.4/W. With the efforts of Chinese PV companies, the cost dropped to USD 1.79/W in 2009, USD 1.035/W in 2011, and USD 0.8/W in 2012.

The situation for international competition is obvious. With leading technologies and invincible large-scale manufacturing, China has formed the skeleton of a technological powerhouse. For the rest, it needs just the flesh and blood—marketable applications for this technology.

There is an argument that the center of global manufacturing is transferring to Southeast Asia where labor costs are much cheaper, while China's demographic dividend is disappearing, and so the cost advantage will vanish correspondingly. I don't think this view is credible.

> I am always wary of dampening expectations for the Chinese manufacturing industry. Without China, there would be no cheap products in the world. Without China's PV industry, the US and Europe would have to spend more taxpayer money to subsidize solar energy development, their new energy would stop developing, and people there would depend more on traditional energy.

The concept of human capital is broader than just the size of the labor force, as it places more emphasis on the "quality" of that labor force. It refers to "the stock of competencies, knowledge, social and personality attributes—including creativity, cognitive abilities—embodied in the ability to perform labor so as to produce economic value. It is an aggregate economic view of human beings acting within economies, which is an attempt to capture social, biological, cultural and psychological complexity as they interact in explicit and/or economic transactions" (Wikipedia). The growth of human capital mainly depends on changes in population, labor force, age structure, urban—rural structure, education, and the labor force participation rate, among other factors.

According to the "Long-Term Growth" Research Group conducted by the Development Research Center of the State Council, China's human capital reached 41.3 trillion yuan in 2012, which was 1.64% higher than in 2011. The ratio of human capital stock to GDP is 1.31:1, which means, 1.31 yuan of human capital stock produces USD 1 of GDP. Among the driving forces of human capital growth, the size of the labor force contributes only a little. In 2012, China's working-age population (ages 15 to 64) was 988.9 million, indicating a year-on-year increase of 0.4%, but its proportion in the total growth of human capital was only 24.3%. That meant the size of the labor force only contributed 25% to the growth in human capital. Meanwhile, the contribution rate of labor force growth to human capital growth is still decreasing. It is estimated that China's working-age population was 992 million in 2013, rising 0.31% compared with that of 2012 and accounting for 20.1% in the total growth of human capital. Its contribution to human capital growth was 20%, which was significantly lower than that of 2012. In 2013, China's human capital is expected to reach 41.9 trillion yuan, an increase of 1.53% over that of 2012.

In general, China's stock of human capital has been growing since 1990. Despite the reduction in working-age population potentially slowing down human capital growth, it will remain a growing trend through increased investment in education and an improving urbanization rate over the next 10 years. Human capital is still important to China's long-term growth.

As for part of the manufacturing sector witnessing "parallel shiftings" (manufacturing shifting to emerging economies and then re-shifting to developed countries), Li Yizhong, Deputy Director of the Economic Subcommittee of the CPPCC, told the media that we must take a realistic approach to analyzing the challenges we are facing. At present, China's growth costs are on the rise and some of this is inevitable, such as labor cost and land cost. So it is quite normal that multinationals shift their production to areas like Southeast Asia, because it boasts lower manufacturing costs. "We shouldn't panic. The quality of China's labor force is better than that of the Southeast Asian countries. We are still competitive in terms of labor force." In addition, a complete industrial chain is being formed in China. With good labor skills and advanced infrastructure, together with honoring and carrying out contracts in an efficient manner, China's manufacturing will come to enjoy new advantages.

The "reindustrialization" proposed by developed countries will challenge and inspire China. China is still in the middle of the industrialization process. It will be time-consuming and arduous for China to realize industrialization by 2020. China must transform and upgrade: shift from the low end to the high end, extend from the manufacturing sector to the service industry, enhance and extend the industrial chain, increase these processes' value and efficacy, get rid of the production mode of processing on order, constantly upgrade and become involved in the high value-adding areas such as designing and branding.

The output of PV products made in China ranks first in the world. It is reported that China's export volume of solar cells and modules accounted for 64.2% of global PV product exports in 2012. More importantly, this large-scale manufacturing is based on high quality and low cost, and this advantage is getting more prominent. Relying on technological advances and operational improvements, the cost of China's PV industry will fall at an unprecedented rate.

Years of exploration have allowed Chinese PV companies to find the right business model for their needs, which is one reason for their low costs. In general, if a company limits itself to the link of cell production, it will suffer from the tight relationship between cost and price. However, if the company's businesses cover the whole industrial chain, it will have more scope to make profits through selling products and providing systematic services such as financing, designing, procurement, construction, grid connection, power generation, power plant operation, and long-term maintenance. In March 2009, QS Solar announced that it would build the country's first universal-type PV power plant in Rudong Port, Jiangsu Province. The thin-film cells used in this power plant would be produced by the company, and the cost would be 60% lower than that of polycrystalline silicon cells.

In the past thirty years or more, the model of "Made-in-China" has endowed the PV industry with cost advantages. Why, then, does the Chinese PV industry lack competitiveness on the world stage?

Core Competency: Technological Advantage

Technology is the core competency in the PV industry. We recognize that China's technological level is low compared to many developed countries. However, this conclusion is not accurate for the PV industry. In PV technology, China is not only a top power in the world, but also an international leader with many key technologies.

Through innovation, technology introduction, technology trade, and corporate M&A, China's PV industry possesses the leading technology in the world. Since Hanergy acquired Solibro, MiaSole and Global Solar Energy, the Chinese PV industry now owns seven thin-film technological lines, of which three have the ability to compete with other world PV leaders, and the other four surpass the US, the Europe and Japan.

Jiangsu Zhongneng Polysilicon, which produces the main raw materials for solar cells, created the new process of "silane-FBR." With this technology, product purity can reach 11N, but the cost is less than half of the modified Siemens process. This means that the Chinese PV industry enjoys more of a competitive edge over the US and Europe, with huge potential for further widening this gap in the future.

When Xi Jinping, then Vice President of PRC, paid a visit to Hanergy's plant where thin-film solar modules based on silicon are made in Shuangliu, Sichuan Province, in August 2011, the plant had just finished construction two months prior. I was there as his tour guide for the plant. I told Vice President Xi that this world-leading production line was independently developed by China. It was also the world's largest silicon-based thin-film cell project. Its annual production capacity could reach 0.3GW, and the photoelectric conversion efficiency of its cells could reach 10%. Xi urged us "to strategically prioritize the strength of independent innovation."

Advantage Element 3: Entrepreneurial Spirit

People are the most important factor in achieving success. China has enjoyed the "demographic dividends" brought by its urbanization, and is proud of its people's entrepreneurial spirit.

Currently, besides American companies still maintaining their strong tradition of having a pioneering and entrepreneurial spirit, companies in Europe, Japan and South Korea have adopted the mode of "CEOs + professional managers." Most of those companies become less innovative, because they are rarely controlled by their founders. Meanwhile, the most noticeable feature of Chinese entrepreneurs is that they are playing the role of "founders + CEOs."

The Mode of "Founders + CEOs"

The past 30 years is a period that witnessed the emergence of the Chinese entrepreneur. It is also a period when these Chinese entrepreneurs created miracles. Facing few opportunities, less demand and no favorable conditions whatsoever, Chinese entrepreneurs nonetheless made painstaking efforts to create opportunities, demands, and conditions. They started businesses from scratch and single-handedly created many economic miracles.

China's PV industry has nearly one million employees, which is the world's largest PV industrial army. Its backbone is the skilled workers and technicians who master the most advanced technology in the world. These technicians include both brilliant Chinese and foreign talent. More importantly, a number of outstanding entrepreneurs are leading this troop. I believe this is one of the key factors contributing to China's leading position in the PV revolution.

The master of economics Joseph Schumpeter highlighted that the function of entrepreneurs is to innovate. He defined innovation as "the setting up of a new production function" that introduces the unprecedented "new combination" of inputs and production conditions into production systems. The rise of a country is often accompanied by the growth of a number of outstanding companies. For example, the rise of South Korea depended on a group of excellent companies such as Samsung and Hyundai.

On August 14, 2013, Michel Sidibe, the under-Secretary-General of the United Nations, visited Hanergy. He told me that his father taught him when he was a child that "A leader must embody three elements: vision, action and heart. Only by doing this, can one achieve success."

> The three essential elements of "vision, action and heart" are especially evident in the entrepreneurs who act as both founders and CEOs. Because the mode of "founders + CEOs" is so popular in China, the country does not lack entrepreneurial spirit in general, nor excellent individual entrepreneurs with pioneering spirit.

Currently, besides the American companies still holding onto their strong tradition of pioneering and entrepreneurial spirit, companies worldwide (chiefly in Europe, Japan and South Korea) have adopted the mode of "CEOs + professional managers." Most of those companies become less innovative, because they aren't controlled by their founders. Meanwhile the mode of "founders + CEOs" is one of the most noticeable features of Chinese entrepreneurs. Hanergy is such a company. With a mode of "CEOs + professional managers," many companies in Europe, Japan and South Korea lack much in the way of pioneering and entrepreneurial spirit, which is a far cry from the mode of "founders + CEOs."

Liu Chuanzhi's Story

China's mode of "founders + CEOs" is a priceless treasure, of which I have a deep understanding. In September 2011, I accompanied the former Vice Premier Wang Qishan to participate in the Sino-UK Economic and Financial Dialogue. Because the meeting could not start on time, the guests had to wait one hour. During that time, I happened to meet the CEOs of a famous bank. To begin with, they looked contemptuous, because they did not know about the Chinese private enterprises attending, like Hanergy. After twenty minutes of talk, their attitude changed and they started to show great respect. Why did their attitude change? For me, it is not because of the large scale of Hanergy, but because those CEOs are professional managers. They lack entrepreneurial spirit and most of them are conservative, wanting the courage and determination to take risks. Their Chinese counterparts have weathered numerous challenges to thrive, and so will continue to move forward through any storm.

Liu Chuanzhi's story is a good example. Known as the godfather of Chinese CEOs, Liu is the entrepreneur whom I respect the most. He is the archetypical founder and CEO.

Mr. Liu was born in April 1944. His paternal grandfather was a native of Zhenjiang City, Jiangsu Province. From 1961 to 1966, he majored in the radar department at the People's Liberation Army Institute of Telecommunication Engineering (now Xidian University) in China. After graduation, he worked at the Commission for Science, Technology and Industry for National Defense. From 1968 to 1970, he was sent to a farm in Zhuhai, Guangdong Province. It was in 1968 that the IT (information technology) giant Intel was founded. From 1970 to 1983, he worked as an assistant researcher at the Foreign Equipment Laboratory in the Institute of Computing Technology of the Chinese Academy of Sciences (CAS). In his own words, he "had researched magnetic recording circuitry for 13 years." From 1983 to 1984, he worked at the Bureau of Personnel of CAS. It was in 1984 that the Institute of Computing Technology began to reform. The researchers there had the chance to go into business. Liu thought this was what he really wanted, so he returned to the Institute.

Liu asked Zhou Guangzhou, former President of CAS, for the two cottages beside the janitor's room, 0.2 million yuan of start-up capital, two partners and the title of deputy general manager. Thus was Lenovo quietly born, largely unnoticed.

With Liu at the helm, Legend Holdings has become a huge enterprise. In 2012, its total revenue amounted to 226.6 billion yuan, with total assets of 187.2 billion yuan, and 60,026 employees. Over those years, Lenovo has established five subsidiaries; Lenovo Group, Digital China, Legend Capital, Hony Capital and Raycom. In 2004, the Lenovo Group purchased IBM's personal computer business and became an international company. Liu believed there were three important elements in managing a company, which are "establishing a team, designing a strategy, and nurturing leaders." With his business ethics, Liu has trained a new generation of celebrity entrepreneurs and made the company a "family business without family members."

I was quite impressed by Liu's comeback.

From 1984 to 2001, Liu served as Chairman and President of Lenovo Group. In 2002, he was the Chairman and President of Legend Holdings, and also the

Chairman of Lenovo Group. In 2004, Yang Yuanqing took the position of Chairman of Lenovo Group.

In the fiscal years 2008 and 2009, the annual loss at Lenovo Group was USD 226 million. When the company suffered losses for eleven quarters in a row, the Group announced in February 2009 that "Liu will return to the chairmanship" to turn the business around. The day after his comeback, Lenovo's stock went up by 11%. According to the market value at that time, the market capitalization of Lenovo had increased by HK$3.3 billion.

During an interview, Liu said, "If your kid is playing on the street and a car is coming, what do you do? Do you worry that you may die in a car crash before you reach him? Lenovo is my kid and I am ready to help him whenever he needs me."

After being Chairman of Lenovo Group again for 33 months, the 67-year-old chairman resigned from his position on the evening of November 2, 2011, but he continued to serve as the honorary chairman and a senior consultant. This shows Liu's far-sighted strategic vision. During the nearly three years since his comeback, Liu always trained and endorsed Lenovo's management, which was led by Yang Yuanqing, and set the tone of Chinese-market-oriented strategy to save the company. However, Liu did not interfere too much in specific operations, leaving enough space for the executives.

Liu said, "Having cooperated with each other for 20 years, Yuanqing and I have become one part of each other in businesses." He believes, "Yang is fully competent to be the chairman and CEO. He will lead the Group to fly higher and to be known by the rest of the world." That night, Lenovo Group announced its second-quarter performance as of September 30, 2011. The data showed that global PC sales of Lenovo had exceeded the average market growth for ten consecutive quarters. For eight consecutive quarters, the company had been the fastest growing company among the world's top four PC manufacturers. With a high market share of 13.5%, it became the world's second-largest PC brand. It was "the best moment" in the history of Lenovo.

Excellent Practitioners in the Chinese PV Industry

Hanergy is a private company adopting the mode of "founders + CEOs." Michel Sidibe, Under-Secretary-General of the United Nations, is the person in charge of global AIDS prevention. He has visited many companies and met with many well-known entrepreneurs, but he was most impressed with his visit to Hanergy. He said it was at Hanergy that he saw people "with vision, action and heart."

Hanergy is undoubtedly forward-looking and strategic. When we built the Jin'anqiao Hydropower Station and joined the team to develop the thin-film solar industry, I had a strong feeling that we could make it.

The construction of the Jin'anqiao Hydropower Station lasted eight years, during which Hanergy faced unprecedented challenges, weathered difficulties, and achieved tremendous growth. The significance of its successful birth goes far beyond the construction itself. I often say, "We should work hard—it is then that God will help us."

The Jin'anqiao project began in 2002. At that time, the United Front Work Department of CPC Central Committee and the All-China Federation of Industry and Commerce were jointly promoting the China Glory Society by organizing private enterprises to inspect Yunnan for further investment. When touring the Jinshajiang River, I learned that there were 100 million kilowatts of hydropower resources waiting to be developed in Yunnan, and the provincial Party committee and government were expecting private capital. I thought this was a historic opportunity, and immediately decided to invest one billion yuan to carry out a feasibility study of hydropower projects in the middle reaches of the Jinshajiang River basin. Our efforts were highly recognized and well supported by the Yunnan provincial government. They soon signed a preliminary feasibility study agreement with us. Later, on September 29, the two sides signed the *Investment and Development Agreement of Jin'anqiao Hydropower Station*. The project investment was 20 billion yuan. Its two phases of construction would have a total installed capacity of 3 million kilowatts, of which 2.4 million kilowatts went to the first phase. The project also became both the first and the only hydropower project with an installed capacity of more than one million kilowatts constructed by a private enterprise.

The necessary investment in the hydropower industry is heavy, and the payback period of capital is long. A project worth tens of billions will not see any economic benefits in the short term. Few private enterprises involve themselves in such a project, and only state-owned enterprises with their strength and governmental support are capable of making such a commitment. For example, the total installed capacity of the Gezhouba Water Control Project is 2.71GW. The project began construction in 1971, spending as much as the nation can bear, and taking 16 years to complete.

Hanergy is such a little-known private enterprise that people would be sceptical about whether it depended solely on itself to build a hydropower project whose total installed capacity is 10% larger than that of the Gezhouba Project. Some said I was crazy. Some thought Hanergy was making a profit by outsourcing the project. And some State Council officials, on the other hand, supported us and gave us an opportunity. Within the company, I eventually persuaded everyone to immediately start work on the Jin'anqiao project. I firmly believed it was the project that I wanted to do. More importantly, I was sure that with our experience in the hydropower business we would be able to make this project a success.

The challenges went far beyond my expectations. The construction lasted for eight years. The heavy investment per day was like a huge millstone, pressing me to death. In order to deal with the 10 million invested per day in the peak periods of the project, Hanergy had to sell its best and profitable power plants built only a few years prior. Those plants embodied the great efforts of Hanergy's best people. Among them the project I was most reluctant to sell was the Nina Hydropower Station in Qinghai Province. The Nina Station had realized grid-connection and generation when Hanergy acquired it with 1.2 billion yuan in 2003. At that time, the financial and real estate markets in China were booming. Many people advised me to sell the Jin'anqiao project and make quick and easy big money. Some powerful enterprises found me and offered attractive prices, but I declined them all. My insistence was not to merely fight for a power station with a capacity of 3GW, but for a principle. This principle was perhaps my faith in clean energy. The idea of "fighting for clean energy" has been ingrained in Hanergy's people's consciousnesses since then.

Persistence demands a high price. In its most difficult times, Hanergy invested all its risk provisions, which were saved up for years, in the Jin'anqiao Hydropower Project. However, the project was like a bottomless pit, impossible to fill. At last, we even borrowed money from our senior managers and their families to maintain investment. In 2007, a ministerial official said to me, "When Hanergy has lived through the challenges of the Jin'anqiao Project, it will be unconquerable." Huang Mengfu, Chairman of the All-China Federation of Industry and Commerce, inspected the project and said, "The Jin'anqiao Project is a miracle of Chinese private enterprise."

Our efforts bore fruit. On March 27, 2011, the hydropower station officially started grid-connected generation. Looking at the towering dam on the Jinshajiang River, the torrents rushing from the dam, and the high-voltage towers on the horizon, I was excited and my thoughts raced.

The construction took eight years, during which I deeply felt the difficulties a privately entrepreneur had to face. I also realized that governmental support was hugely important for our private enterprises. I even found that private enterprises were highly relevant to the country's future. Serving the country with new energy and contributing to people's welfare have become two of Hanergy's missions.

Liu set the Chinese-market-oriented strategy for Lenovo Group, and Chinese PV companies did the same. They stood at the strategic intersection of choosing the road, the market, their companies should take. The Chinese market is a magic key to the door that a large fortune hides behind.

The majority of PV entrepreneurs are the owners of their companies. They have the following characteristics: first, they grew up in the context of China's reform and opening up, so most of them are revolutionary, open-minded and far-sighted. Second, most of them are the founders of their respective businesses. They start businesses from scratch, oversee them growing strong, and experience various hardships. They become indomitable and dauntless in coping with various situations. Third, their success is closely related to national policies and the country's development. Thus, they have a strong desire to serve their communities and their country. They take the idea of "strengthening the country and enriching the people" as their corporate mission.

Chinese entrepreneurs do not lack courage, strategic vision or innovative spirit. They need to defeat themselves and resist temptation. Here we quote a metaphor, "climbing Everest," from Liu Chuanzhi. Liu always says he is living in a period when China lacks business culture, but he sketches out an Everest for himself and conquers it by his willpower alone. With strong determination, he has time and again reached each of his targets until he finally realizes his ambition.

> In just one decade of the 20th century, China's PV industry rose rapidly in the world, which justifies the role of entrepreneurship. China's leading entrepreneurs will play a greater role and show even more strength and fortitude in the future revolution of new energy and on the journey to realize China's leading position in the global PV industry.

China's PV Companies

Over the past 10 years, if there is an industry whose halo is as bright as the one around the Internet, it must be the PV industry. If there is an industry that creates wealth as much as the Internet does, it must be the PV industry. If there is an industry whose ability to attract capital can compete with that of the Internet, it must be the PV industry.

But time flies when you're having fun. The international PV anti-dumping and anti-subsidies duty investigations caused by the crazy expansion of PV capacity became the straw that broke PV's back.

"Crisis" in Chinese means both "danger" and "opportunity"; that is, crisis represents opportunity. Reality has told us that the road of expanding production capacity and the mode of "having raw material imported and products exported" do not work anymore for the Chinese PV industry. This crisis is a perfect opportunity to change from following to leading, from capacity expansion to technological progress, and from disorder to healthy development.

The First Generation of PV Enterprises: The Rise and Failure of Crystalline Silicon

It was reported by *Huaxi City Daily* that on March 20, 2013, the Wuxi Intermediate People's Court ruled Suntech to be bankrupt and so had to be reorganized, according to the *Enterprise Bankruptcy Law*. Similarly, the situation of LDK, an upstream company, was not particularly bright. However, it is undeniable that Suntech and LDK are two of the most vigorous companies representing this sunrise industry. They are known as the "two PV heroes." These two companies were founded by Shi Zhengrong and Peng Xiaofeng, respectively, who are of the new generation of entrepreneurs. Shi and Peng with their companies played a model role in the earlier stage of the PV industry. In just a few short years, China's PV production capacity accounted for more than half of the world's, and among the global top-ten PV module producers, Chinese producers take the first five positions.

It was reported that when Shi Zhengrong, a 37-year-old doctor, returned to China from Australia in 2000, he arrived at Wuxi with only a laptop and a business plan of no more than a few pages. According to Hong Yuqian, general manager of Wuxi Venture Capital Group Company, Ltd., Shi presented his business plan to Hong and several officials from the local technology bureau in a governmental guesthouse near the Zhongshan Road in October 2000. "We felt the plan was risky. Domestic power demand was not high, and power generation was in overcapacity in 2000. We found there was no domestic demand for solar power generation. In addition, we knew nothing about the overseas markets. We couldn't know if it was a promising business plan."

Hong, former Director of Wuxi Machine Tools Company, Ltd., had been a manager for ten years and later served as Deputy Secretary of Wuxi Municipal Finance Bureau. In 1999, the MOST launched a project to promote technological innovation, and Wuxi was one of the pilot cities. The Wuxi municipal government funded the establishment of Wuxi Venture Capital Group Company, Ltd. under its administration. Considering the financial operation risky, the local government appointed the just-retired Hong as the general manager. The collaboration with Dr. Shi was the first challenge for this company.

Hong took the business plan to the senior technicians of Wuxi China Resources Huajing Micro Co., Ltd for advice. This company was the largest manufacturer of integrated circuit chips in China. Its upstream raw material was silicon materials. Ten days later, Hong got the answer: it was a good high technology project, so it was commercially viable. But there was one uncertainty, that it might be a little early to invest immediately. Hong asked them, "Would you mind an investment with us?" They said they were not interested for the time being. In fact, many companies were not brave enough yet to engage in this sector.

In January 2001, Shi convinced the Wuxi municipal government to support his project. According to disclosed information, the registered capital of this company was USD 8 million, of which Shi accounted for 25%, and the technology shares accounted for 20%, equivalent to USD 1.6 million in total; the cash dividend accounted for 5%, equivalent to USD 0.4 million; the other USD 6 million was contributed by eight other shareholders, namely, Wuxi Venture Capital Group Company, Ltd., Wuxi High-Tech Venture Capital Co., Ltd., Wuxi Keda Innovation Investment Ltd., Wuxi Guolian Trust & Investment Co., Ltd., Little Swan Group, Wuxi Mercury Group, Wuxi Shanhe Group, and Shanghai Baolai Investment Management Ltd. The last five companies were corporate shareholders, among which the Shanghai Baolai Investment Management Limited had not invested and its shares were re-distributed among the other shareholders according to stock equity.

In September 2002, Suntech's first solar cell production line of 10MW was officially put into operation. Its production capacity was equivalent to the sum of the whole country in 1998. Its operation had narrowed a fifteen-year PV gap between China and the world.

However, Shi did not stick with his thin-film strength, although he had been in charge of R&D in Pacific Solar Pty., Ltd. for more than six years. In light of China's advantage in manufacturing, he turned to producing crystalline silicon cells.

In its initial stage, Suntech did not fare too well. According to the magazine *China Entrepreneur*, the sales of Suntech in 2002 were more than 10 million yuan, and the losses were over 7 million yuan. The situation worsened in early 2003: in order to get bank loans, the company had to mortgage all its property; and its start-up team fell apart. In August 2004, the third production line at Suntech started production as the company was suffering a severe downturn and even its equipment was pledged to the construction team.

Luckily, several major state-owned shareholders provided loans to Suntech in turn. In 2003 and 2004, Wuxi City also approved nine projects funded by the government for this company. The financial support totaled 37 million yuan. All that helped Suntech out.

It was in the month when Suntech's third production line started operation that this unpleasant situation changed. The German government revised its

Renewable Energy Law to subsidize solar energy and other new energies. Germany became the first country to consume new energy since the enforcement of the Kyoto Protocol. Germany's demand for PV cells was boosted instantly and caused the 2004 global PV market to realize a year-on-year growth of 61%.

When most PV companies did not have enough appetite to swallow this piece of cake, Suntech suddenly saw an opportunity to reap the benefits from this new market. In 2004, its net profit reached USD 19.80 million. In the first three quarters of 2005, the net profit totaled USD 20 million. This time, Shi was quite keen and captured the subtle changes in the PV cell market. Even lacking funds, he was still brave enough to expand his production and eventually created his own success story.

On October 19, 2005, Suntech went public on the NYSE, raising USD 400 million. In the same year, Shi became "the richest man in China."

Perhaps it was the demonstrative effect of Suntech making profits or possibly it was the increasing overseas demand that ultimately made PV the most prosperous industry in China. The PV industry quickly became regarded as the only high-tech industry in which China can compete with the rest of the world. In addition to Suntech, a group of celebrated companies has emerged. Some Chinese provinces and autonomous regions, such as Jiangsu, Hebei and Inner Mongolia, have prioritized the development of the solar PV industry. Nationally, there are at least ten PV companies with a scale similar to Suntech. PV producers, little and big, amount to nearly one thousand. The development of these companies is uneven, but their total capacity is amazing. Globally, little newly installed solar power capacity is annually about 29.79GW, while the PV module production of China in 2011 was 30GW, which is almost equal to the total global amount.

Facing the rapid development of China's PV industry, many people seem to neglect a simple fact: the development of the PV industry relies on demand from outside. Once the US and the European markets change, and governments substantially cut off their support for the PV market, the export of Chinese PV products will inevitably suffer a tremendous hit.

When China's PV industry is in trouble, "the disadvantage of being first" will become the Achilles' heel of the PV frontrunners. This is a capital-intensive industry. In order to gain advantage from being first, radical frontrunners are prone to borrow heavily to expand capacity. With this rapid expansion, the cost of equipment and raw materials is unexpectedly low.

When a Chinese polysilicon manufacturer commenced commercial production of polysilicon in 2008, the cost was USD 110/kg. When the company started its second phase of production in March 2011, the cost had dropped to USD 60/kg. If the cost could maintain itself at a level above USD 400/kg, front-runners reasonably spend more than the latecomers do, because the former will make profits some day in the future. However, if the cost falls to USD 40/kg before prior investments are recovered, that spells big trouble. It is widely recognized that the cost of re-establishing LDK would be less than its debt.

This disordered and uncontrollable capacity expansion caused the international PV anti-dumping and anti-subsidies duty investigations, which became the straw that broke the PV Helios's back. This was a predictable crisis, but the first-generation Chinese PV companies were not well prepared to deal with it. Facing the shrinking European market, those companies feel that it has become difficult to make progress.

The Second Generation of Solar PV Companies: Realizing the "Thin-Film Dream"

However, Chinese solar PV companies have not stopped exploring the industry's potential. Based on the experiences and lessons of the first generation, the second generation of solar PV companies such as Hanergy and ENN blazed a trail in both the technology roadmap and the business model, and so created a new world for themselves.

It was in 2009 that Hanergy entered the PV industry. At that time, crystalline silicon technology was relatively sophisticated, while thin-film technology was still in development. Many insiders were optimistic about the crystalline silicon PV industry. Some even took it for granted that PV *is* crystalline silicon. Therefore, choosing a technology roadmap became a problem.

After careful consideration, Hanergy's decision-makers decided that, as the solar PV industry is a strategic emerging industry, their decisions must be based on the nature of this industry. "Emerging" means the business is extremely promising. "Strategic" means it concerns both general and long-term interests. In the long run, thin film and flexibility are the development direction of PV technology. PV products based on crystalline silicon technology are energy consuming, heavy in weight, and rigid. Those shortcomings are difficult to overcome. PV products based on thin-film technology are energy saving, lightweight, as well as flexible, and perform well in weak light. In the future, solar energy will be inevitably applied to distributed power plants and various mobile power generation devices. Undoubtedly, the merits of thin film are more obvious. As for the photoelectric conversion efficiency and cost of thin film, these can be solved technologically. Moreover, the major business of Chinese PV companies was crystalline silicon-based production when Hanergy wanted to enter the field, so there was no need to become yet another face in a crowded industry. Due to lack of strength, some followers have to compete with products or services similar to others. Their outlook is certainly bleak. Thus, Hanergy decided to find a new way by employing the Blue Ocean Strategy. It resolutely chose the thin-film roadmap, which still got the cold shoulder in China.

Looking back, our choice was correct. The investigations imposed by the European Commission and the US were directed at crystalline silicon products rather than thin-film ones. For Hanergy, that oversight seemed like a stroke of luck. More importantly, thin-film technology did indeed develop as rapidly as expected. Within a few years, the conversion efficiency of thin film modules has increased to 18.7%, which is the same as, or even higher than, that of crystalline silicon, while the production cost is much lower.

Technological leadership is the key to developing PV enterprises. It is often said that excessive production capacity poses risks. In fact, it is not being large scale, but being large scale without enough of a market that poses risks. Products with low technology and high cost will have no market, so the key is technology. Having the most advanced thin-film technology in the world, Hanergy is braver and more confident than ever in rapidly expanding its capacity.

In 2012, the British *Financial Times* and the American *Wall Street Journal* reported Hanergy's overseas acquisitions twice within five months. Reuters even published an interview (an article of about 5,000 words) with Zhou Jiesan, an executive of Hanergy Group, who was in charge of Hanergy's overseas acquisitions. Those frequent reports from renowned international media outlets helped Hanergy attract more attention, both at home and abroad. The PV tide was going out and so provided a good opportunity for Chinese enterprises to pick "pearls." "MiaSole's deal with Hanergy would be the latest example of a US solar startup being rescued by a larger Asian industrial manufacturer." Those comments had a strong flavor of Western style. The thing they cared about was not the little-known Chinese buyer from afar, but the two shining "pearls"—Solibro and MiaSole.

At home, once the state-run Xinhua News Agency and the *21st Century Business Herald* reprinted those foreign reports, the domestic media and their websites flocked to report on Hanergy. This company came under the spotlight again.

The domestic media chose several points to highlight: What are these two companies? Why does the international media pay so much attention to them? What is the purpose of Hanergy's acquisitions against a backdrop of anti-dumping and anti-subsidies duty investigations imposed by the EU and the US?

Abiding by relevant regulations and their contracts, Hanergy did not disclose any information to the public. The domestic media mostly just cited the afore-mentioned foreign reports. Several media outlets only based their articles on the views of individuals in order to analyze Hanergy's motives. In contrast, the Reuters reports were more internationally focused. The agency noted, "The purchases also give Hanergy technology the ability to compete with First Solar Inc., the biggest maker of thin-film panels."

First Solar was the then-current leader of the global PV industry. In 2009, it replaced Q-Cells to become the world's largest PV company. In 2012, its mod-ule production capacity reached 2.5GW. First Solar is a thin-film-based company like Hanergy. Globally, it is the only company beside Hanergy that realizes the falling costs in the thin-film business. Cadmium telluride technology belongs to a branch of thin-film PV technology. First Solar has this technology and enjoys a technological monopoly with Antec Solar, a Germany-based cadmium telluride thin-film cell manufacturer. The conversion efficiency of cadmium telluride cells in tests has reached 17.3%, and the results have been independently confirmed by the US-based National Renewable Energy Laboratory (NREL). Some even say First Solar is America's "national treasure."

Hanergy treats First Solar as its most important competitor. In 2009, Hanergy decided to set a production target of 3GW for the end of 2012. If suc-cessful, it would surpass First Solar. At that time, more than half of its senior employees and executives did not believe Hanergy could make it, because the largest production capacity in China was only tens of megawatts. In December 2012, Hanergy announced that its target of 3GW had been realized.

So, what is the impact of Hanergy's acquisitions (Solibro and MiaSole) on First Solar?

The first impact is the conversion efficiency. Solibro is the world-record holder in the conversion efficiency of thin-film solar cells. Its small-sized winner CIGS cell achieved the world's highest efficiency of 18.7%. Of course, this is only a test result in a laboratory, which is to say, mass production cannot currently be achieved at such high conversion efficiency. So, what is the efficiency for mass pro-duction? Solibro has realized an efficiency of 14.7% in mass production. When the efficiency of solar cells is improved by 1%, the cost of solar power generation is lowered by roughly 8%.[3]

Solibro is a venerable company established in Sweden. Sweden is the research and development center of Europe. It spends 4% of its GDP (as opposed to 1.83% in China) on research, the highest among European countries. Solibro's founders all graduated from Uppsala University, the oldest Nordic university and a paradise of learning. Imagine being neighbors with knowledge itself, and you will know the value of this university. CIGS technology was born in the Ångström Laboratory of the university in 1983. The leader of this technology was Lars Stolt (the incumbent CTO of Solibro), who therefore is known as the "Father of CIGS thin-film solar energy." Later, with his colleagues at Uppsala University, Lars Stolt founded Solibro, which was acquired by the world solar giant Q-Cells in 2009.

The American company MiaSole is also a powerful competitor for First Solar. The conversion efficiency of its CIGS thin-film solar modules is 15.5% in mass production.

Located in Silicon Valley, MiaSole ranks first among the 200-odd local new-energy companies in terms of technical indicators and technological strengths. The members of its R&D team mostly come from Intel, the global giant of the computer industry. They embody the changes in Silicon Valley, where an energy revolution has replaced the information revolution, and are witnesses to the transformation of Silicon Valley into "Sun Valley." John Doerr is the primary investor in MiaSole, and he is called "Bill Gates" by the investment community. He once invested in the computer giant Compaq and the Internet giants Google and Amazon. In the rising star of Silicon Valley, MiaSole, he also invested USD 550 million.

Why did Hanergy acquire two CIGS companies at the same time? The answer is "strategic consolidation of global technologies." Hanergy selected 20 leading thin-film PV companies from the US, Europe, Japan, Israel, and mainland China, and conducted site visits. First Solar was among them and was ruled out at last, because its cadmium telluride technology emitted the toxin cadmium. Although relevant recycling technology exists, it did not fit with Hanergy's concept of clean energy. Finally, five companies with different techniques became the final options for Hanergy's acquisition. Although both Solibro and MiaSole adopted CIGS technology, the former used evaporation techniques, and the latter adopted a sputtering technique. Having a variety of techniques has two advantages: One is that diversified products can provide a wide range of application scenarios. For example, Solibro's products are more rigid and MiaSole products are more flexible. The other advantage is that sharing patents can greatly promote technological progress. Currently, technology research and development is relatively conservative in Europe, the US and Japan. For example, First Solar exclusively monopolizes cadmium telluride technology. Meanwhile, the various techniques mastered by Hanergy can supplement and promote each other in an open environment. That is why we need to consolidate global technologies.

Maybe this is why the celebrated Western media paid such great attention to the two acquisitions. Actually, they knew little about Hanergy, which came from far away in the East. If they had known that the production capacity of Hanergy had reached 3GW at the end of 2012, they would perhaps make different interpretations about its acquisitions.

Drawing lessons from the first generation of PV enterprises and their mode of "having raw material imported and products exported," Hanergy adopted a business mode of "building a high-end, large and complete product line." "Being large" means production on a massive scale. In fact, Hanergy has consecutively established nine production lines for thin-film cells, and acquired three foreign companies. In addition, its production capacity in thin-film cells has come to be ranked first in the world in just a short time. Some would cast doubts, perhaps asking whether it is merely "blind expansion." In fact, Hanergy has enough strength to do so, and it is permanently and strategically significant for this company to promptly seize opportunities and to expand capacity.

The PV industry is a marriage between the high-tech industry and the energy industry. The nature of the high-tech industry is its large-scale production. To lead in technologies requires increased investment in R&D, while the investment cannot be realized without financial strength. However, without large scale, there can be no financial strength. The leading high-tech companies are generally large enterprises. In 2009, when Hanergy was purchasing high-end equipment, I met with two enterprises. They were the Switzerland-based Oerlikon and the US-based Applied Materials. When they were told that Hanergy expected to realize a capacity of 3GW, they thought I was joking. Because First Solar, the world's largest American PV company, had a production capacity of less than 2GW at that time. Those two companies *did* have advanced technology, but their costs were hard to cut due to their small scale. Later, one was purchased and the other left the thin-film field.

The nature of the energy industry is also its large-scale production. The energy industry covers every corner of society, so the sales of its products have to be massive. Energy products are basic commodities in economic and social development, so their price should not be as high as those of some consumer goods or luxury goods, and their benefits must come from "economies of scale." The successful enterprises on the world stage are mostly large or super-large enterprises.

Therefore, since the PV industry was born as the result of the marriage of these two big industries, it must be large for both its survival and its development. This is why Hanergy increased its production capacity to 3GW as its first phase.

For Hanergy, this is only an initial pattern. Its aim is to reach a capacity of 20GW in 2015. Many people may think this is an astronomical figure, but in reality it is possible to achieve. Actually, the generation capacity of solar power plants is only equal to one-quarter of traditional power plants, so a solar installed capacity of 20GW is only equal to a hydropower installed capacity of 5GW. The current hydropower installed capacity of Hanergy has reached 6GW. If Hanergy can achieve the goal of 100GW in installed PV capacity by 2020, and if its thin-film PV production capacity accounts for one-quarter of the world's, then we can say it's big.

"Being complete" means a business covers the whole industrial chain. Unlike many solar PV companies' businesses, Hanergy's business extends from the production of cells and modules to the entire PV industrial chain. The upstream high-end equipment manufacturing, the midstream thin-film cell production, and the downstream power plant construction together constitute Hanergy's "business model of the whole industrial chain." Its business covers raw material exploration, technology R&D, equipment development and manufacturing, system designing, as well as the construction, management and operation of power plants. This kind of business model can give full play to Hanergy's strengths in R&D, technology, manufacturing and engineering. It also can settle the conflict between photoelectric costs and business benefits.

Recently, knowing that Hanergy's production capacity has reached 3GW, people will ask, "The market can consume such a large amount of capacity?" They *do not* know that Hanergy is essentially a power generator in nature. Its profits do not simply rely on manufacturing, but mainly depend on system solution services, solar energy applications and power generation.

Since its HeYuan Manufacturing Base (in Guangdong Province) finished construction on July 1, 2009, Hanergy has always synchronized its construction of solar power plants and production bases. The annual capacity of 3GW *is not* in surplus.

As early as 2010, when Hanergy had only been a part of the PV industry for two years, it incorporated Hanergy Holdings (America) Limited in San Francisco, and built two small-sized solar power plants, which have been in operation ever since. Those moves can be regarded as a springboard for Hanergy to enter the US market.

Coincidentally, another silicon-based thin-film company, ENN, signed a letter of intent with the State of Nevada at the China–US Economic and Trade Cooperation Forum in February 2012. This company intends to invest USD 5 billion in the state to build an eco-center of clean energy, including PV power bases and PV module manufacturing bases.

With its business model the whole industrial chain, Hanergy has promoted solar energy applications in various ways. In Italy and Greece, it provides distributed generation services for some commercial projects. In China, it tries to cooperate with famous businesses. For example, Hanergy and IKEA jointly launched a "green home strategy," in which Hanergy will provide thin-film modules of 0.383GW and build distributed power plants for IKEA-owned Chinese stores and its 67 suppliers over three years. Once put into operation, those systems of distributed power generation will meet 15% to 20% of the annual electricity demand of the Chinese stores and all the electricity demand of IKEA's Chinese distribution centers.

Previously, Hanergy's household solar power generation system sold at IKEA UK had enjoyed popularity. I once paid a home visit to a British customer of mine. He told me that he spent 4950 pounds installing 25 pieces of thin-film panel (a 3kW solar power system), which can generate 3600kWh per year.* As a reward, he gets an annual subsidy of 576 pounds from the government and saves 261 pounds on his electricity bill (half of his own generation can meet his power demand). In addition, his profit on electricity sales is 81 pounds, for the surplus electricity is sold to grid companies at 0.045 pound/kWh. In total, he earns 918 pounds every year, equivalent to an annual ROI of 13%.

There is a long-standing dispute between "integration strategy" and "specialization strategy" in the PV industry. The "specialization strategy" focuses on one link and making it strong and superior. The "integration strategy" consists of the "horizontal integration strategy" and the "vertical integration strategy." Horizontal integration is a consolidation of many firms that handle the same part of the production process, while vertical integration is typified by one firm engaged in different parts of production. The mode of the whole industrial chain is a kind of vertical integration. Theoretically, those strategies have their strengths and weaknesses. The key is whether strategies and enterprises "match." According to its circumstances and understanding of the PV industry, Hanergy chooses the vertical integration strategy, which is a result of seeking a good "match," and being "different" rather than making a statement that this is the only correct mode.

A business covering the whole industrial chain has its own issues, its own pros and cons, and risks. The most distinctive characteristic of this business model is that the links interact with each other within the whole value chain, and all depend on downstream sales for ROI, so it takes a longer time to see a positive cash flow. Without continued financing, the business is likely to give up halfway. Meanwhile, this business model requires companies to know precisely the status quo and trend of every link in a timely manner. Moreover, companies must be capable of appropriately handling challenges to prevent any mistake in the links from affecting the whole chain. The mode of the whole industrial chain indeed fails in some Chinese companies. Most are caused by lack of financing and making mistakes in some of the links.

---

* Editor's Note: 1 pound is about 9.97 yuan (exchange rate in January 2014)

Hanergy is making great efforts to achieve its goals. The three goals are intrinsically and organically related to each other. High technology can achieve large scale; large scale and great strength can support technological R&D and upgrade technology; and a business covering the whole industrial chain helps to expand the scale, to improve efficiency, and to ensure Hanergy soars "higher" and grows "larger."

The World Looks to China for PV

2013 will be the turning point in the PV application market in China. From this year onwards, China will completely shake off the label of being a "big nation in PV manufacturing, but small in application," and transform into a powerful nation in both PV manufacturing and application. At present, China has been making mid- and long-term plans for PV development, and the disclosed figures are big enough to cheer up the PV industry. China's PV industry will play a leading role in the world PV industry and then in the world energy revolution.

Leading the World in PV Application

China has been the world's largest producer of PV products since 2007, and its share in the world market is rising year on year. Even after the US and the EU launched the anti-dumping and anti-subsidies duty investigations and raised market thresholds, China's ever-expanding PV manufacturing advantage has not been lost. China's PV cell production capacity reached 21GW in 2012, accounting for 63% of the world total. If we include Taiwan's production capacity, this proportion is 73%, and so we can see that China truly lives up to its reputation as a big PV manufacturing nation.

2013 is the turning point because it's expected that China's newly installed capacity for PV power generation this year will be the largest in the world for the first time. In the past, 90% of China's PV products were exported. However, since the PV feed-in tariffs scheme was introduced in 2009, the PV application market has been rising rapidly, and installed capacity in 2012 reached 8.2GW. According to market expectations, Germany's newly installed capacity in 2013 will drop from 7.6GW to 5GW. In China, the National Energy Administration announced at the beginning of 2013 that it expected China's installed capacity to reach 10GW. Judging from the current situation, China's capacity will at least reach 6 to 8 GW, overtaking Germany and ranking first in the world.

Change always runs faster than plans. As for changes in the PV application market, China made many rather forward-looking development plans in the past, but eventually it turned out that we had underestimated the growth rate of the market. In July 2013, the State Council issued *Several Opinions on Promoting the Healthy Development of the Photovoltaic Industry*, which will significantly raise the total installed capacity in the PV market to 35GW. This means that in the following 3 years, there will be 10GW of newly installed capacity on average every year in China.

The government's latest plan will improve China's share of PV installed capacity in the world dramatically. By contrast, the capacity of the European market is plummeting. According to data from HSBC, the share of the European market in world total installed capacity will decrease from 68% in 2011 to 32% in 2013, and 25% in 2015. Meanwhile, China's share will increase from 11% in 2011 to 27% in 2013, and remain at 23% in 2015, almost equal to that of Europe. China and Europe will become the two joint-largest PV application markets in the world.

At present, related institutions, entrusted by the National Energy Administration, are helping to draw up the PV development plan for 2030 and 2050, which will serve as guidance for decades to come. According to some sources, China's PV capacity will reach at least 100 GW in 2020. This means that with the 35GW capacity installed before 2015 excluded, the average annual installed capacity will reach 15GW between 2016 and 2020. Between 2020 and 2050, there will be a substantial increase in domestic capacity. By then, China's long-term planning for annual PV installed capacity in 30 years may be raised to 30 GW, which means the newly installed capacity per year will be equivalent to 30 units of 100 million kilowatts of thermal power units.

According to future energy demand forecasts, by 2030 China's total energy demands will reach 7.5 billion tons of standard coal, and then renewable energy will play a huge role in substitution. According to the plan, by 2050 PV power generation capacity will reach 1 billion kilowatts. 1 billion kilowatts is equivalent to the current total installed power capacity in China. At the end of 2012, China's total installed capacity was 1.14 billion kilowatts, which means that by 2050, these 1.14 billion kilowatts can be completely replaced by PV.

According to China's existing reserves of resources, PV mid- and long-term planning also shows the necessity of large-scale PV development. With economic development, China and many countries in the world face the dual pressures of energy and economy. China is now the world's largest energy consumer, largest electricity consumer, largest importer of coal, and largest carbon dioxide emitting country. As a major energy consumer, how much conventional energy do we have in reserve? As mentioned earlier, China's reserves of coal and natural gas are only enough for the next 33 years, oil for less than 10 years. This means that in 30 years' time, we will have no coal, oil or gas to use. For the sustainable development of China's energy and for the sake of the environment, PV power generation and renewable energy are the only way out. China's energy transformation must be completed in the next 20 to 30 years. China will advance from the current energy mix of conventional energy sources accounting for 80% of total energy, to one in which renewable energy becomes dominant.

China's national policy encourages PV development, and there is a strong industrial base for domestic PV manufacturing. So it is not only necessary but also possible that the breakthrough in China's PV application market is just around the corner. Now many countries are hesitating in their national PV policy, and the market is in decline. So it will be easy for China to dominate the market, and lead the global PV industry and even the new energy revolution.

The Engines of Technology and Manufacturing

The explosive growth of China's PV market is not only a solar and energy revolution, but also a revolution in investment. China lists new energy as one of the seven new strategic emerging industries because of its enormous economic potential. Not just the new energy industry, but also associated emerging industries, including energy saving and environmental protection, high-end equipment, and new materials. They also have a bright future. According to current development prospects and estimates, investment in the PV industry will surpass that of any other industry. And if it can substitute for fossil fuel energy, its investment scale will outpace the total fixed assets of the existing industries by a large margin, and will promote sustained and rapid development in the next 30 to 50 years.

As a result, we will not miss this opportunity to take the lead in world PV development. We should improve our anticipation of the market, and continue to pay attention to upgrading the whole PV industrial chain. Technology and manufacturing is the foundation for China to becoming a PV major power. Different from oil, natural gas, coal and other fossil fuel energy, it is impossible for anyone to monopolize solar energy. The basis of competition lies in technology, and the one who has the core technology can dominant the market.

The past decade has witnessed an annual growth rate of 50% in global PV power generation, a development pace that no industry has achieved before. And China's PV industry has made great contributions to it. Wolfgang Palz, president of the American Council on Renewable Energy, said that PV module manufacturers in China have made irreplaceable contributions to the world. Chinese enterprises have made great efforts to improve production efficiency and manage quality. This has helped to lower the cost and price of PV modules by a large margin within a limited amount of time, improve the competitive edge of PV power generation against conventional power, and add momentum and capacity for the rapid development of the market. Without Chinese enterprises, the world PV industry would not have achieved its current status, and the 800,000 jobs in the industry would not have been created.

At present, the manufacturing cost per watt for a PV module has fallen to 4.5 yuan, and the per-watt manufacturing cost of PV systems has also been lowered to between 8 and 9 yuan. This has enabled the cost of PV power generation to decrease by 80% within just a few short years. The PV-generated price of electricity was 4 yuan per kilowatt-hour in 2006, but now it is less than 1 yuan. These are all great achievements by Chinese PV enterprises.

These achievements could not have been made without relevant technological advances. PV development is based on the manufacturing industry and high technology is at its core, both of which are indispensable. At present, China has the advantage in both, especially in manufacturing. Although there is overcapacity in PV manufacturing, it is a temporary and purely structural problem. Recently, the government has listed many industries that have overcapacity, but the PV industry is not among them. This suggests that the overcapacity in the PV industry is not to be taken at face value. Overcapacity here is merely temporary. There appears to be overcapacity now, but that capacity may not be able to satisfy the domestic market in the future. According to the mid- to long-term plan, the average installed capacity between 2020 and 2050 will reach 30 GW. So if we include demand from the international market, PV production capacity may be far from enough.

There is another reason why PV overcapacity is structural. Due to the development of distributed PV power generation, domestic demand for thin-film batteries will increase substantially, which makes the thin-film battery actually in short supply, with only crystalline silicon cells in surplus.

> Currently, the world's PV technology is also gradually changing from the crystalline silicon era to the thin-film era. Thin film has become the mainstream of the world's PV power generation. In this regard China has grasped the opportunity, which is also why we are confident in the future development of the PV industry. The one who wins in thin film will also be able to win in the general future PV competition. We are not only able to be the best in the field of crystalline silicon, but also in the more advanced thin film.

By the end of 2012, Hanergy's production capacity for thin-film solar had reached 3 million kilowatts, surpassing First Solar and becoming the global leader in the thin-film industry.

Hanergy's surge out of the PV downturn will not only enable China to achieve leapfrog-style development in new energy technologies, but also to make strategic contributions to the global film industry, especially along two technical routes: silicon-based and copper indium gallium diselenide thin films. Without Hanergy, we might have lost the world's most advanced CIGS technology and silicon-based thin-film technology. Through technology acquisition, Hanergy achieved both the large-scale production and upgrading of these two technologies in China.

Through the integration of global technology, Hanergy now possesses seven product technologies, including amorphous silicon–germanium, amorphous silicon–nanometer silicon, copper indium gallium diselenide, etc. The highest conversion rate of the small size champion module in CIGS technology has reached 18.7%, and thus is a global leader in thin-film solar industry. Hanergy continues to integrate global thin-film technology: it has again chosen five companies from many candidates as technical acquisitions. If it realizes all these acquisitions, Chinese companies will continue to lead the world in new energy technologies in the future.

Hanergy has the most advanced thin-film technology both in China and around the world, via independent intellectual property rights. Its leading position can be seen by considering the following factors: first, less consumption of raw material; second, short energy payback time; third, the small temperature coefficient of thin-film modules, and good low-light efficiency; fourth, it can be made into flexible film modules; and fifth, product diversification potential.

Hanergy's leadership also means that China is taking the lead. Now Chinese film solar technology has overtaken traditional Western PV powers. If the government can clarify its position that thin-film technology—with a variety of advantages including environmental friendliness, flexibility, and a wide range of applications—is the right direction for solar industry development, and support the thin-film PV industry in terms of capital, technology, markets and other areas, China will have more of a competitive edge, and be better qualified to lead the development of the world PV revolution.

NATIONAL STRATEGY: THE RIGHT PATH FOR CHINA'S PV INDUSTRY

Introduction

China's PV industry is still facing a series of problems: overcapacity, difficulties in integration, dependence on raw material import and final product export, the financial strain, and loose grids. We still have a long way to go to solve these problems.

In terms of our mindset, we should adopt an overall, long-term, and flexible vision of the global economy, and adjust the way we face globalization accordingly. In line with a people-oriented and environment-friendly perspective, we should understand the significance of new energy development—especially the development of the PV industry—for scientific development. We should use "innovative thinking" instead of "inertial thinking" to seek solutions for energy problems. We should not focus on immediate benefits without considering strategic benefits. We should elevate the new energy revolution to a proper strategic position. We should always maintain a positive attitude, give full play to the advantage of China's new energy industry, and firmly implement the catch-up strategy.

In practice, the government's actions need to be "in place" in four areas: strategy in place, planning in place, policy in place and measures in place. Only in this way, can the "three-step replacement" strategy (growth, incremental, and subject replacement) to replace traditional fossil fuels with new energy be fully implemented, and our dream of changing the world with clean energy achieved.

Now, all we need is opportunity.

Behind The Imbalance of the PV Market

When you review the structure and status of China's PV industry today from the perspective of the industrial chain, you find that a competitive industrial system has yet to be formed. We can say that China's PV industry is still in the "Warring States Period."

Eight problems are particularly prominent: overcapacity, difficulties in integration, dependence on raw material import and final product export, a slow start in the domestic market, the urgent need to strengthen the grid support, the mainstream crystalline silicon technology, the structural adjustment in overseas markets, and corporate financial strains.

We must first analyze these problems in order to solve them.

Production Capacity Out of Control

According to the statistics of the Global New Energy Development Report (2013), the world's newly installed capacity in the PV industry has reached 30.3GW, a record high. However, the world's production capacity of crystalline silicon solar cells was as high as 60.1GW in 2012, for which China's production capacity accounted for 38.7GW (close to two-thirds of global total capacity). In other words, even if all 2012 PV orders were for crystalline silicon products supplied by Chinese companies, there would still be overcapacity in China's PV industry.

The major reason for the current overcapacity in China's PV industry is the excessive expectations of enterprises and local governments for the market.

In 2010, the global PV industry finally recovered from the 2008 global financial crisis. That year the newly installed capacity of the global PV industry increased to 16.8GW, with an increase rate of up to 127% compared with 7.4GW in 2009. In 2011, the newly installed capacity of the global PV industry increased by 65% to 27.7GW. According to a May 2012 EPIA report, in the most optimistic of scenarios (with governments implementing effective policies to stimulate the PV market), the world's new PV capacity could reach 40.2GW in 2012 and 77.3GW in 2016. However, if we do not consider policy drivers, EPIA's predictions for newly installed capacity in 2012 and 2016 are as low as 20.2GW and 38.8GW respectively.

There is an implicit premise in the EPIA's two market predictions for the PV industry, namely, that the development of the PV industry relies on policies.

In 2010, in the wake of the global PV industry rebound, China's PV entrepreneurs found more optimistic signs in the market. It is worth noting that China's PV industry has followed a "bottom-up" path since the very beginning. In the early stages, the whole industry was dominated by scientists and entrepreneurs supported by local governments, lacking effective national overall planning.

On the one hand, current capacity is but a momentary surplus. It is in large part due to fewer overseas orders and insufficient domestic demand. Once the market environment changes, the surplus will disappear. On the other hand, the current overcapacity in China's PV industry is mainly due to polysilicon cell production, while the production capacity of thin-film batteries is inadequate.

Figure 5–1 Predictions on global installed PV capacity between 2011 and 2016.

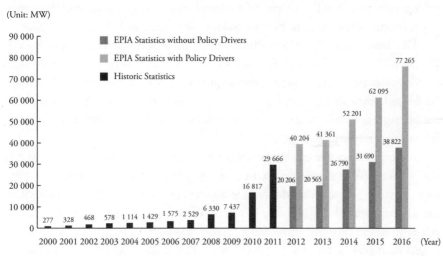

(Unit: MW)

Source: EPIA

After 2010, local governments craved output value, profit and tax, and employment. The industry faced more pressure from in-depth industrial transformation and upgrading. Entrepreneurs and financial institutions had great expectations for the industry. Thus China's PV industry blindly expanded.

Statistics from the Zhejiang Solar Energy Industry Association show that from September 2010 to March 2011, 78 new PV companies emerged in the province of Zhejiang, an average increase of 13 per month.

This phenomenon is not unique to Zhejiang. According to *China Energy News*, by December 2011, 31 provinces, municipalities and autonomous regions had issued preferential policies to support the PV industry. More than 300 cities actively participated in developing the PV industry. Among them, more than 100 proposed building large-scale PV industrial parks, so as to form several industrial clusters centered on the Bohai Bay, Yangtze River Delta, as well as the Southwest and Northwest regions of China.

Take the province of Zhejiang as an example. The production capacity of PV enterprises in Zhejiang in the first seven months of 2011 was over 2.3 times that of the total production capacity in 2009. According to the statistics of NDRC, by the end of 2011, production capacity for crystalline silicon solar cells in China had reached 40GW, nearly twice that of 2010 (21GW). In other words, the current crystalline silicon cell production capacity in China was achieved before 2012.

Since 2010, the fast development of China's PV industry ultimately led to a problematic situation. There were PV enterprises everywhere, but most of them were not competitive enough, and their surged production capacity outpaced demand.

On September 19, 2010, Li Junfeng, then Deputy Director of the Energy Research Institute of the NDRC, asserted at the New Energy Conference held in Wuxi, Jiangsu, that, "China's production capacity grows too fast, so it is inevitable that there will be surplus in this PV market next year." Although many people agreed with him, they were cautiously optimistic, as nobody wanted to miss the opportunity at hand.

At the same conference, Qu Xiaohua, CEO of Canadian Solar Inc. (CSI) said, "There will be no surplus in the new energy industry in the medium to long term, [though] there is imbalance in the short term, there is still much room for the development of this industry in the next five years." At that time, CSI's solar cell production capacity witnessed a rapid expansion from 420MW in 2009 to 720MW. It planned to reach 1.3GW in 2012. At the head of the Chinese PV industry at the time was Suntech Power, headquartered in Wuxi. Its production capacity surged from 1.1GW in 2009 to 1.8GW in 2010, making it the largest PV cell producer in the world.

However, the global PV market did not follow China's expectations. According to the *Global New Energy Development Report* (2013), PV capacity in the world only expanded by 30.3GW in 2012. That is 25% (nearly 10GW) lower than the EPIA's 40.2GW. The anti-dumping and anti-subsidies duty investigations launched by the US against Chinese PV products, the shrinking European market as well as its subsequent launch of similar investigations are the three external factors which led to the deterioration of China's PV industry starting in 2011.

Of course, we need to make a specific analysis of the current overcapacity in China's PV industry.

Difficult Integration

Several months after the production capacity of the Chinese PV industry reached its historical peak in May 2012, the CIConsulting Industry Research Center (hereinafter referred to as "CIConsulting") concluded in the *In-Depth Analysis on China's Solar PV Industry from 2012 to 2016 and Development Planning and Consultation Report* that, "there may be a large-scale integration in the PV industry next year (2013)." The report stated, "[...] in the early stage of changes, most enterprises can manage to keep going with their accumulated wealth, but over time, it is difficult for some to survive because of liquidity squeeze. It can be said that this was only the beginning of the most difficult days for PV enterprises, the second half of this year (2012) and next year (2013) will be the most difficult period for PV enterprises, and the real challenge has just begun."

This assessment was proved correct by national policies six months later. On December 19th, 2012, a State Council executive meeting chaired by then Chinese Premier Wen Jiabao identified the main problems at that time as severe overcapacity, over-reliance on external demand, and operating difficulties. The meeting also decided on the policies and measures to promote the healthy development of this industry. In order to solve the serious overcapacity issue, the State Council clearly stated that, on one hand, China should speed up industrial restructuring and technological progress, use market forces to encourage corporate M&As, eliminate backward facilities, and improve technology and equipment; on the other hand, we should regulate the development of this industry, actively explore the domestic PV market, and improve policy support. It also made clear that we should give full play to market mechanisms, reduce government intervention, and prohibit local protectionism.

> By the second half of 2012, the central government and PV entrepreneurs realized that they had to absorb severe overcapacity through reorganization. However, the only large enterprise that has been reorganized so far is LDK. Its tortuous reorganization and the dominance of local governments in that process showed that reorganizing the PV industry is still difficult in China.

Rumors of LDK possibly going bankrupt shocked the entire industry in China in early 2012. However, against the PV industry's overcapacity backdrop, especially that of the global PV crystalline silicon industry, it is not difficult to understand the predicament the world's largest poly-silicon film producer found itself in. The released information showed that its total assets amounted to USD 6.637 billion, total liabilities to USD 5.962 billion, while asset–liability ratio reached a record 89.82% as of March 31, 2012.

After February 2012, as the LDK crisis gradually spread out, there were more rumors of reorganization. Possible restructuring parties included Pingmei Shenma Group, Yingli Solar, GCL, Jinko Solar, Sinoma, CECEP, China National Building Material Group, China Guodian Corperation, China Reform Holdings Corporation (often nicknamed the second China Investment Cooperation), as well as other SOEs in Jiangxi, such as Jiangxi Copper Corporation.[1]

The released information shows that on October 22nd, 2012, LDK announced that it would sell 19.9% of its floating shares to Hengrui New Energy Co., Ltd, whose shares were held by the local government. Hengrui is a newly established company engaging in solar energy, investment and related businesses.

60% of its shares are owned by China Hi-Tech Wealth Group Co., Ltd; the other 40% by Xinyu State-owned Asset Management Company. To outsiders, LDK's reorganization reflects the will of the Xinyu government, and could even be perceived as nationalization.

The released information shows that in 2005, the Xinyu government raised 200 million yuan to support Peng Xiaofeng in establishing LDK, which was considered a major feat at the time. A few years later, the city government received large returns. In 2010, LDK had become one of the world's largest silicon producers with a capacity of 3GW, and its sales exceeded yuan 20 billion. In the same year, the economy of Xinyu enjoyed leapfrog development. Its revenue increased from 3 billion yuan in 2008 to 5 billion in 2009, and again to 8 billion in 2010. Its GDP soared to 60.5 billion yuan in 2010 (or 2.1 times that of 2005 when LDK had just been brought into production), with an average annual growth of 15.5%.

In 2011, the output value of new energies centered on PV accounted for more than half of the total for the three pillar industries in Xinyu, namely new energies, new materials and steel. In the past, the steel industry had been the sole pillar industry of Xinyu, had contributed the most revenue and created the most jobs. In 2011, LDK paid 1.36 billion yuan in taxes, becoming Xinyu's largest contributor. That year, Xinyu's total fiscal revenue was 11.13 billion yuan, an eight-fold increase compared with the year of LDK's foundation.

Yet, after the LDK crisis, in the first half of 2012, Xinyu's supersized industry generated profits of only 1.607 billion yuan, a decrease of 52.7% year on year. In fact, the economy of Xinyu had a symbiotic relationship with LDK. With an expected production value of 150 billion and a planned investment of 500 billion, LDK was not only important for the upstream and downstream enterprises in the established PV industrial chain, but also for tens of thousands of jobs and the economy of Xinyu.[2] Therefore, the local government was unwilling to see LDK decline or go bankrupt.

It was reported that in February 2012, when LDK faced the risk of liquidity squeeze, the municipality of Xinyu launched a 20-billion yuan development stabilization fund to provide financial support for LDK. In June, in order to repay a maturing 500-million-yuan trust loan, the Xinyu government even used its municipal coffers to provide guarantee and helped LDK renew the loan for another three years.

However, saving LDK is an arduous task. The local government's financial support was only a contingency plan. In the long-term, well-established strategic investors must be introduced to reorganize LDK.

There were a number of strong potential strategic investors who were said to have cooperated with LDK. In the end, LDK partnered with Hengrui New Energy, 40% of whose shares are held by the Xinyu State-owned Asset Management Company, even though it only brought LDK 22.93 million dollars.

This tortuous process is yet to be revealed to the outside world. Nevertheless, for Xinyu City and the province of Jiangxi, there are at least a few advantages. The land-mark enterprise LDK remains in Xinyu, Jiangxi. In the future, if LDK re-emerges, it can continue to contribute huge revenue to the local government and create many jobs. More importantly, based on LDK, Xinyu may be able to establish the PV industrial chain, with upstream and downstream enterprises, with an expected pro-duction value of 150 billion and a planned investment of 500 billion.

However, in my opinion, in that reorganization, market mechanisms weren't given full play. We need to see whether they can phase out the backward produc-tion capacity.

> The anti-dumping and anti-subsidies investigation launched by the EU and the US in 2012 harmed China's PV industry, but if we use that pressure to motivate industrial restructuring, it could have a positive result.

The ongoing bankruptcy reorganization of Suntech might be a more market-oriented example of restructuring in China's PV industry. In August 2013, Ren Haoning, a PV industry researcher of CIConsulting, told the press that there were over 100 PV enterprises above designated size in China, but that through further M&A, there would be three to five giant enterprises left in the end.

Importing Raw Material and Exporting Final Products

Many people wonder how LDK, the largest supplier of silicon pieces, fell to the brink of bankruptcy in one year. In my opinion, one important reason is that China's PV industry relies on importing raw material and exporting final products.

Take the development pattern of crystalline silicon PV as an example. At pres-ent, we grossly process raw materials, ship them overseas for fine processing, and ship them back again to China to make cells and modules before selling the final product abroad. Chinese companies are the PV industry's workshops. According to founder of Acer Group Stan Shih's "smile curve," the most lucrative sections of the entire value chain are the two ends (R&D and marketing). Enterprises with no R&D capability can only serve as agents or workshops, and earn little money for their hard work. For enterprises with no marketing capacity, even excellent products are dumped as waste when outdated. Thus only by continuously moving toward the high-added-value section can a company maintain sustainable development and operation.

Figure 5–2 The smile curve

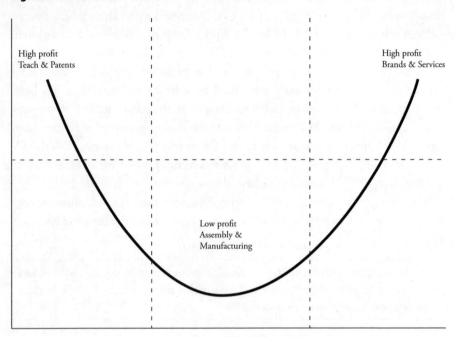

High profit
Teach & Patents

High profit
Brands & Services

Low profit
Assembly &
Manufacturing

Under these circumstances, the industrial chain layout is inherently improper. After having acquired manufacturing capacity, Chinese enterprises, as market followers, should aim at moving towards the two ends of the industrial chain. At this moment, it is the most important task in China's PV industry upgrade. In retrospect, we can find a clear track in China's PV industry, which started at the processing and manufacturing section in the global PV industrial chain. It is now time to make a difference in research and marketing.

Since 2007, China has ranked first in the production of PV cells. In 2009, it contributed 40% to the global output. However, high production does not mean that China is a giant in this industry. For example, China is the largest iron and steel producer in the world, yet Chinese enterprises have little say in the pricing of iron ore. China is the world's largest automobile producer, but there is no Chinese equivalent for Volkswagen or Chrysler. Before 2011, hidden behind the prosperity of the Chinese PV industry lay the plight of the industrial chain and the helplessness of enterprises.

Take crystalline silicon PV for example. There are six links in its industrial chain: solar-grade silicon materials, silicon ingots, wafers, cells, modules, and system integration. In the whole chain, the preparation and making of crystalline silicon wafers is the most technically demanding link, while manufacturing cells and modules is the easiest. Especially in the modules manufacturing link, the competition in cost is severe, creating risks for companies.

However, Chinese PV companies are mainly cell and module manufacturers, while seven manufacturers in Europe, America and Japan monopolize crystalline silicon material. In 2008, they controlled more than 70% of the global supply of poly-silicon. Europe is the largest PV generation market. In 2008, its installed capacity of PV systems accounted for nearly 80% of the global total. In this industrial structure, China, a so-called "PV giant," is merely a "major manufacturer," or, to be more accurate, a "major cell and module manufacturer." We must acknowledge that in such a new industry, China has not yet shed its role as the world's workshop. Strategic high ground and high added-value sectors such as capital, technology, and markets are controlled in other countries, while low value-added manufacturing sectors in the downstream were introduced in China.

Such deficiencies cause very serious consequences. Due to China's dependence on raw material import and final products export, its PV industry is trapped with limited bargaining power and high vulnerability.

> In general, due to such dependence, it is difficult for China's PV industry to form a complete industrial chain. China may end up with lopsided development, and Chinese enterprises, especially large ones, are bound to be subject to the influence of others. This situation must be changed, and it is changing.

In the upstream, foreign manufacturers, represented by the seven crystalline silicon suppliers, have a very strong bargaining position compared to numerous other cell and component manufacturers. In 2005, the explosive growth of the PV industry caused a global shortage of polysilicon, making its price soar. In 2004, solar-grade silicon materials were sold at only USD 30 per kg; in 2005, due to a sharp increase in market demand, the price increased to over USD 40 per kg; in December 2007, the price was as high as USD 400 per kg; and in September 2008, it even reached a historical high of nearly USD 500 per kg. It shows that profit in the upstream is very lucrative. The one who owns silicon is king in this sector.

Many Chinese companies rushed into silicon material production not because they intended to upgrade the structure of the industry but in pursuit of profits. Investment in silicon material production boomed as a result. As of 2010, nearly 50 Chinese companies planned to build or were building or expanding poly-silicon production lines. The total production capacity then was over 100,000 tons, and the total investment was more than 100 billion yuan. During this period, China's silicon capacity took up less than 1% of the global market in 2006, and its share increased all the way up to 8% in 2007. By 2008, its polysilicon production capacity accounted for 37% of the global total.

However, influenced by the outbreak of the global financial crisis in 2008, the global PV market shrank, and silicon material prices also dropped. Starting in 2009, silicon material prices dropped all the way from USD 300 per kg to about USD 60 per kg by the end of 2012. Many Chinese enterprises involved in large-scale investment before 2010 were also caught in the predicament.

In the global PV industrial chain, though the "who owns silicon is king" situation no longer existed after the outbreak of the financial crisis, we cannot overlook the importance of the upstream (the crystalline silicon preparation link). The upstream dominates the entire PV industrial chain with its technology and capital advantage. In fact, the ability to obtain a stable supply of silicon material and wafers has always been one the main concerns for Chinese battery cell and component manufacturers.[3]

The Unbalanced Market

The Chinese PV industry is more reliant on overseas markets than on overseas material. Unfortunately, this situation has not fundamentally changed.

Figure 5–3 The price movement of spot poly-silicon in the world (2012)

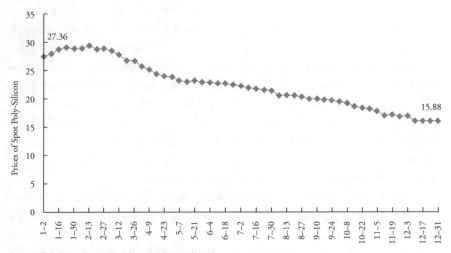

Source: Global New Energy Development Report (2013)

When China's PV industry began a large-scale expansion of its production capacity, the US and the European Union launched one investigation after another against Chinese PV products.

Media reports said that Suntech, who had the largest market share in the US, was the biggest victim of the investigations. At that time, Zhang Jianmin, head of the Investment Relations Department at Suntech, said to the media that the US market absorbed 25% of Suntech's total sales. However, after the US published the results of its investigation, a 31.22% anti-dumping tax was imposed on Suntech. Zhang Jianmin admitted that Suntech could not earn a profit with such high tax. Suntech's first quarterly report in 2012 showed that its operating revenue was USD 409.5 million, its gross profit was only USD 2.4 million, while it paid a punitive tariff as high as USD 19.2 million. Suntech's net loss was USD 133 million dollars.[4]

Figure 5–4 Top 10 countries in cumulative installed PV capacity in the world (2012)

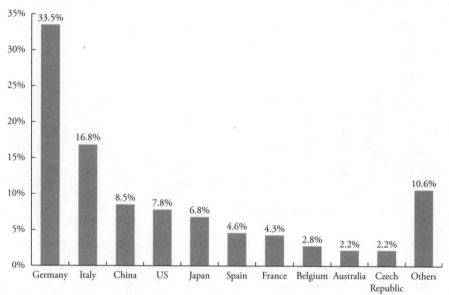

Source: Global New Energy Development Report (2013)

The dependence of China's PV industry on overseas markets is a well-known and often discussed fact. Although China's domestic PV market has accelerated development in recent years, the world's major markets are still in Europe, the US and Japan. As of 2012, the top 10 countries (in size of PV market) accounted for 89.4% of the global cumulative installed capacity. In regional distribution, the European cumulative installed PV capacity reached 66GW, 68.3% of the global total. In other words, on average, before 2012, nearly 70% of all PV products produced in China were exported to Europe. 8% were exported to the US, 7% to Japan, and only 8.5% were consumed in the Chinese domestic market, and most of that after 2010.

But after the global financial crisis in 2008, the Spanish market declined sharply. Especially after 2010, Germany reduced its policies supportive of the PV industry, which directly resulted in a slowdown in the global PV market. The first victims were Chinese PV companies, because most of them were manufacturing components and cells, a link in the chain where the technical threshold is lowest and the competition is fiercest.[5]

Figure 5–5 The distribution of newly installed capacity in the world (2012)

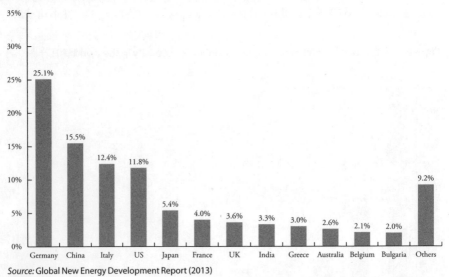

Source: Global New Energy Development Report (2013)

Subsequently, the overseas market also slowed down, the trade war escalated, and the overcapacity problem in China became more obvious. Expanding China's domestic PV market was a necessity.

In 2012, China's PV industry met with more difficulties triggered by the investigations launched by the US and the EU. The root cause was the imbalance between the domestic and the overseas markets. To be more specific, domestic demand was not fully unleashed. In that year, China produced about 40GW of PV cells, while China needed less than 5GW. That was the crux of the problem. Restructuring the market and expanding domestic demand was not just a contingency plan, but also an imperative and strategic measure for developing China's PV industry and the new energy revolution.

Difficulties in Grid Integration

During the NPC & CPPCC in March 2009, Liu Hanyuan, Member of the National Committee of the CPPCC, became famous for submitting a proposal entitled "*Issuing Consumption Coupons to Everyone to Stimulate Domestic Demand.*" He proposed granting consumption coupons worth 4,000 yuan to every citizen each year. Another of his proposals, "*The PV Industry Shall be Covered by Domestic Policies on Demand Expansion,*" also had a far-reaching impact.

To reduce the cost of PV generation and popularize PV in China, Liu Hanyuan suggested that we should learn from European practices, and ask the State Grid to purchase all the PV power generated at the subsidized on-grid tariff. The central government should set the initial subsidized tariff at between yuan 1.5 and 2 yuan/kWh, and adjust the tariff once every 3 years or 5 years. Meanwhile, the *Million Solar Roofs Plan* should be carried out in areas with favorable conditions, and solar PV generation systems installed on family roofs. Low-interest loans or even free-interest loans should be provided to the participant families to purchase generation systems. In addition, the government should purchase the family-generated power at the subsidized tariff, so as to produce extra income.

Liu Hanyuan made such a proposal at the NPC & CPPCC, not only just because he is a member of the National Committee of the CPPCC, but also the board chairman of the Tongwei Group. This group is one of the biggest aquatic feed producers in the world, as well as one of the biggest PV enterprises in the province of Sichuan. As early as 2007, the Group had invested 5 billion yuan to establish a poly-silicon manufacturing plant, the Sichuan Yongxiang Co., Ltd., in the municipality of Leshan.

The status of power grid enterprises was notably demonstrated by Liu Hanyuan's explicit proposal.

Indeed, the growing domestic demand for PV power is not only the result of bigger solar power plants, but also of all kinds of building-distributed power systems. However, all PV development needs support from the State Grid; otherwise this is nothing but empty talk.

In the future, in order to achieve smooth grid integration, we have to strike a balance between PV stations and the grid, between PV power and fossil-fueled power, between power generation and supply, as well as among various shareholders. There are technical issues, as well as challenges in operation, management, policies, and institutions. The new smart grid is a combination of technologies and economics, of development and reform. Therefore, we need great effort and firm resolution to build it.

In recent years, the "large-scale ground-mounted PV station" has become one of the hottest terms in the PV industry. PV enterprises that have overstocked modules seem to take the on-grid tariff and local governments' craze for attracting business and investment as their life-saving straws. At the end of 2011, in order to enjoy the NDRC's on-grid tariff, which was set at 1.15 yuan/kWh, many PV enterprises rushed to build PV power stations on the Gobi Desert of Golmud, in the province of Qinghai. Nevertheless, due to lack of communication with grid companies in the preliminary stages, many PV power stations were not linked with the grid, and remained idle for a long period.

When these PV stations, in which hundreds of millions of yuan were invested without hesitation, were set up, they frequently encountered difficulties with grid integration. The troublesome and trivial application procedures of grid companies would take the stations half a year to a whole year. The changing requirements of local grid companies also led to cost increases.

Procedures are the first challenge faced by PV enterprises. According to the *Guidelines on the Integration of Newly-built Power Units into the Grid* published by a certain grid power transaction center, during the preliminary scheduling of a PV station, PV enterprises have to submit data and materials, and fill in application forms at least 50 times in various offices. It takes as long as an estimated 4 or 5 months to complete all these procedures.

Moreover, even if they are approved, it is hard for PV enterprises to cover their extra costs. For example, a large amount of grid-integration equipment often not included in the original costs must be taken into account—such things as stability controllers, reactive compensators, active power controllers, and reactive power controllers. The expenditures for grid-integration equipment, as estimated by the manager of a PV enterprise, can be as much as 5 or 6 million yuan, no matter how small the power station is.

At least at this point, it is not easy for PV enterprises to deal with grid companies. In addition to rising costs, they have to handle affairs beyond their professional scope. They even have to communicate with local residents by themselves if they need to acquire land for building transmission lines and transformer substations. However, they have little leeway in bargaining with grid companies, because grid companies have the final say on integration matters.

An insider has complained: "Is it possible to set a unified standard for the specific expenditure of grid integration across China? Even if it is impossible due to various local circumstances, is it possible to make the procedures open and transparent?"[6]

In fact, even national demonstration projects, such as the "Golden Sun Program," have difficulties in grid integration. At the beginning of 2012, after the "Golden Sun Demonstration Program" had been in place for 3 years, a nationwide investigation showed that only about 40% of its units had been linked to the grid. *China Energy News* reported that the biggest resistance to grid integration comes from grid companies. The "Golden Sun Program" encourages users to supply surplus power to the grid and to consume the power generated by them. Their shortage can offset the power generated from the grid companies. In this way, users purchase much less power from grid companies, cutting the grid companies' performance. Moreover, the Golden Sun policies are advocated by the Ministry of Finance, but poor coordination with grid companies in the preliminary stage of implementation has resulted in more difficulties.

According to another insider, grid companies are worried that grid integration will have a negative impact on their performance. They remain dubious about the quality of PV power. In addition, grid companies are unwilling to bear the additional costs caused by the installation of reactive compensation equipment in the process of grid integration.

When it comes to the difficulties of new energy integration, including PV power, Wang Sicheng, a researcher at the Energy Research Institute of the NDRC, has pointed out that the biggest challenge lies in the lack of explicit policies rather than in the reluctance of the grid. Shi Lishan, Deputy Director of the Department of New Energy and Renewable Energy, NEA, believes on the other hand that "we need time to improve integration between new energies and conventional energies. In recent years, as the government and enterprises have updated their knowledge of integration, the State Grid is trying its best to tackle the difficulties in this process." In addition, many people have gotten used to our centralized management of large-scale power stations, which is very different from distributed solar power generation.[7]

Crystalline Silicon Is Not the Only Choice

The issues under discussion are mostly related to crystalline silicon. I would like to stress that crystalline silicon is not the only choice in the PV industry. As mentioned above, when we talked about over-capacity in the PV industry, thin film is not a relevant material, because its production capacity is not excessive.

As far as I know, in current large-scale ground stations, there is no big difference between crystalline silicon application and thin-film application. Many Chinese enterprises still employ the first-generation technology of crystalline silicon, which has some insurmountable defects, especially in BIPV. At the same time, the conversion efficiency of thin-film PV power has been improved through technical upgrades. Its efficiency is equal to or even superior to that of crystalline silicon, while its cost is lower.

In September, 2011, Zhang Gong, Professor at Tsinghua University and a famous expert on solar cells, expressed an idea at the Forum of Environment-friendly Energy and Green Industry. According to him, if we want to lower the huge costs of solar cells, we have to improve the conversion efficiency of PV power and reduce manufacturing costs. Thin-film cells are made of more inexpensive silicon and less material than poly-silicon cells. Moreover, they consume much less energy than poly-silicon cells. Therefore, the use of thin-film cells is the trend in solar cell development.

Why do most thin-film PV enterprises in the world choose to walk such an untrodden path? It is because they firmly believe that thin-film cells have advantages over poly-silicon cells, including low costs, mature technologies and less pollution.

Jiang Qian, Chief Researcher for New Energy Industry at China Investment Consulting said, "Compared with crystalline-silicon cells, thin-film cells equipped with upgraded technologies have much more potential for reducing cost. It's possible to enlarge the production capacity of thin-film cells to cover 20% of the whole solar energy industry. There are many physical methods to achieve breakthroughs in thin-film cell technologies, such as the use of an ion beam to deposit nano crystalline silicon films. If they could improve their methods, domestic enterprises would appear much more promising."

The cost advantage of thin-film cells is obvious, compared to crystalline-silicon cells, and thin-film cells are also good substitutes for thermal power and other conventional energies. At the end of 2011, a manager of ENN Solar Energy spoke highly of developing thin-film cells in an interview with the media. He also predicted that, as the proportion of the PV industry in the global energy industry gradually increases, thin-film cells, promoted by demand for application in large-scale PV stations, would achieve rapid development.

Why are crystalline silicon cells still playing a leading role in the global PV industry? The *Global New Energy Development Report (2013)* reveals that the capacity of the world's crystalline-silicon cells in 2012 reached 60.1GW (38.7GW produced by China), while the capacity of thin-film cells in the same year was as low as 7GW. So why do many crystalline silicon enterprises choose not to get into the thin-film business? The reasons are simple. First, they do not have the technologies; second, to achieve equivalent capacity, they have to initially invest in thin films as much as 8 to 10 times more than for crystalline silicon; third, many of them aim to gain quick profits, and are not really optimistic about the PV industry. The earlier crystalline silicon rush is also an important reason for the severe overcapacity in crystalline silicon cells in China.

Xiao Jinsong, APAC Channel Director of Applied Materials (Asia-Pacific), had comments on the advantages and disadvantages of crystalline silicon and thin-film technologies: "I admit that crystalline-silicon cells have some advantages. Actually, both kinds have their own pros and cons." In his opinion, the generation efficiency for crystalline silicon cells is higher with higher costs, while that of the thin-film cell is lower with lower costs. Thin-film cells can be applied on a large scale in deserts or areas where land resources are abundant or temperature difference is large, because they are highly applicable to low light and adaptive to temperature differences.

In my opinion, a greater advantage of thin-film cells is their wide applicability. Thin-film cells can be processed into sub-transparent and flexible cells, which can be widely applied for military use, civil use and BIPV.[8] In recent years, with the fast growth of thin-film cells, many problems (such as low conversion efficiency and raw material shortages) have been solved thanks to technological advances and expanded industrial scale.

Up to now, many people, including some insiders, held that the conversion efficiency for thin film is lower than that of crystalline silicon. As a matter of fact, the efficiency of thin film is already equal to or even higher than that of crystalline silicon. In 2012, mass production conversion efficiency of thin-film modules produced by Hanergy reached 15.5%, while the average conversion efficiency of domestic crystalline silicon modules was only about 15%.

In the future, there will still be big opportunity for improving thin-film cells, which will become mainstream globally. In Silicon Valley in the US, all companies are engaged in R&D on flexible thin film instead of crystalline silicon.

China should give priority to second-generation technologies, namely thin-film technologies. China should fully support the PV industry, equip it with cutting-edge technologies on a considerable scale, and allow it to be dominated by thin-film solar power. We should select several thin-film producers for pilot programs, and ensure that they are global leaders in core technologies in order to realize leapfrog development in China's energy sector.

The Cake China Lost?

As far as I can tell, the investigations launched by the US and the EU did not stop China's PV industry from reaching the international PV market. PV enterprises, industry organizations and the central government of China could expand the market by building better relations with foreign enterprises and the international market.

A piece of good news is that 95 Chinese enterprises and the European Commission, under the coordination of the China Chamber of Commerce for Import and Export of Machinery and Electronic Products, have reached an arrangement for prices in order to settle the trade disputes over PV products exported from China. Consequently, the market shares of Chinese PV products in the EU market have become smaller, and the price advantage has been impaired.

Nevertheless, 60% of the market has been maintained, and vicious competition among Chinese enterprises avoided. In a sense, the investigations may have had some positive repercussions.

> In the future, with thin films, China could re-occupy the market, which was lost during the investigations by the US and the EU; and this market is large. In 2011, the shares Chinese enterprises enjoyed in the EU market reached 11.4GW (worth 99.18 billion yuan), while shares in the US market reached 1.1GW (worth 9.57 billion, yuan). The shares in the two markets totaled 108.75 billion yuan.

Figure 5–6 Production capacity and output of thin-film cells between 2004 and 2012.

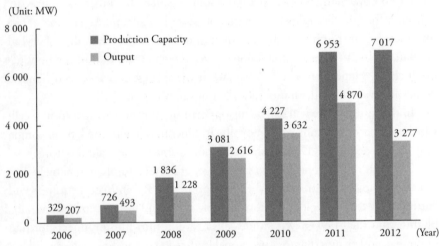

(Unit: MW)

Source: Global New Energy Development Report (2013)

The Capital Chain Crisis

Since China's PV industry as a whole was trapped in crisis in 2010, most companies blamed external factors, including shrinking foreign markets, trade barriers, slow response in the domestic market, overcapacity of the entire PV industry, credit crunch, etc. However, after assessing these factors objectively it's necessary for us to identify the companies' own problems.

Over ten years, in the leapfrog development of China's PV industry, some burgeoning world-class enterprises have been cultivated together with some successful entrepreneurs. However, each individual in the PV industry should be acutely aware that it is industrial opportunities more than the endeavors of entrepreneurs that lead to success. Business can prosper only if Chinese entrepreneurs in the PV industry continue to improve their internal strengths.

Affected by the sluggish global market and the investigations, the entire PV industry in China had been swamped. As a result, some long-term problems within Chinese PV enterprises were exposed. For example, the financial issue was most prominent in 2012.

An enterprise bears huge financial pressure if it doesn't have smooth fund flow. It has been reported that the top 10 largest enterprises producing poly-silicon products bore accumulated liabilities of 110 billion yuan, with an asset–liability rate of more than 80% in 2012. The root cause of such a plight is that many of them unwisely pursued fast development with a large amount of loans. However, financial institutions will tighten the money supply once the industry declines abruptly.

At the beginning of 2012, there were a lot of media reports on the LDK crisis. I believe that it was financial problems which led to their difficulties.

In a report from the *21st Century Business Review*, Li Hua, director of the Risk Management Department of a Chinese commercial bank, said that the roots of the crisis lay in LDK's industrial chain expansion and Peng Xiaofeng's preference for capital-intensive projects. "Because LDK Solar always invests in capital-intensive projects, it was not sensitive to the efficiency of capital use. Other Chinese PV enterprises which mostly started in business by producing cells and modules always have had strict budgets, so they usually give up on huge investments like that of LDK."

The capital-intensive strategy adopted by Peng Xiaofeng was once successful in the silicon business. According to a senior employee of LDK, "LDK took the lead in the silicon business during 2007 to 2009 largely because of its equipment. We bought out 70% of suppliers' capacity. It meant that any other enterprise could only acquire 30% of suppliers' capacity at most. In other words, it never could catch up with LDK in those years."

In June 2007, LDK showed a more unswerving faith in capital-intensive expansion after being successfully listed on the New York Stock Exchange. Then it launched its Industrial Chain Expansion Strategy. As a result, the poly-silicon project, which received the most investment, became the bringer of financial difficulties.

In August 2007, LDK announced that it was beginning to establish two poly-silicon projects whose total capacity could reach 16 thousand tons. The investment in fixed assets reached up to 13 billion yuan. At that time, the price of polysilicon soared from USD 40/kg in 2005 to USD 300/kg, and peaked in September 2008 at USD 475/kg. Afterwards, affected by the global financial crisis, the major markets for PV generation slackened off. Additionally, the capacity of polysilicon failed to be unleashed. The price plunged to about USD 60/kg when the first assembly line with an annual capacity of 5,000 tons was put into production.

In July 2012, the price glided further down to about USD 22/kg, while production costs were as high as USD 38/kg. Since capacity was mostly established when the price peaked, LDK Solar lost more than 10 dollars per kg in silicon production. We could even say that LDK Solar's 13 billion yuan in the polysilicon project was an invalid asset, while the net assets of LDK Solar were merely about 4.2 billion yuan.

The statistics in the LDK Annual Report show that LDK's fixed assets continued expanding from 2007 to 2011, by USD 0.337 billion, USD 1.697 billion, USD 2.609 billion, USD 2.993 billion and USD 3.872 billion per year, respectively. Meanwhile, its asset–liability rate also went up, by 47.09%, 76.58%, 80%, 81.43% and 87.68%, respectively. The manager pointed out: "LDK invested too much. Though it had a large amount of fixed assets, it could not recover cash in such a recession. That was the fundamental cause of its debt crisis." [9]

Figure 5–7 Fixed assets and debt-to-assets ratio of LDK Solar

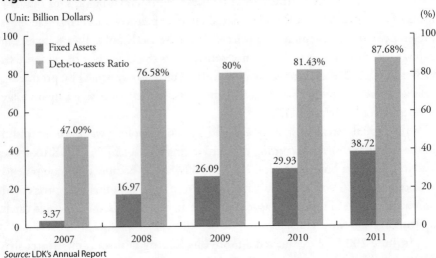

Source: LDK's Annual Report

Many Chinese PV enterprises have encountered financial difficulties since 2011. We learn from incidents and acquire more knowledge about the PV industry: the PV industry is a technology-intensive and capital-intensive industry; it requires a huge initial investment and a long wait for payback.

We learned lessons from the capital chain crisis. On one hand, PV enterprises will be aware of its features, and take measures to avoid a breakdown of the capital chain in operation. On the other hand, financial institutions also should be aware of these things, and regard the credit support for them as a long-term blue chip rather than a short-term product. This emerging industry could really benefit the nation and its citizens if PV enterprises and financial institutions learn from the past.

Strategizing is The Way To Go

Today, ordinary people in China are familiar with globalization. However, in the past, it appeared as an unattainable prize for Chinese industries.

For many Chinese entrepreneurs, it is a bitter story. China sacrificed its domestic market in exchange for technologies, and subsequently became the world's factory. Nevertheless, the new energy revolution guided by the PV industry might be able to transform Chinese entrepreneurs' participation in globalization.

Therefore, we have to change our thought pattern. As we become more and more people-oriented and environmentally-conscious, we will be able to develop new energies, especially the PV industry, and put "scientific development" into practice. We will be able to tackle energy problems only if innovative thinking takes the place of inertial thinking. To promote the energy revolution, strategic interests should outweigh immediate interests. It is possible for China's new energy industries to catch up with and surpass others as long as we give full play to our late-arriving advantages.

To Globalize in Another Way

A strategic vision means looking at the overall situation, as well as considering long-term development and changing circumstances. To have insight means to be able to detect the true nature of things through superficial phenomena.

With such a strategic vision, we could solve the issue of global resource integration. This means that any nation or enterprise could integrate the resources they need worldwide. This observation is, however, not enough. In fact, nations want to maximize their benefit through the integration of resources.

Such a vision could allow us to better understand the nature of the investigations. The US and the EU purchased our crystalline-silicon cells because they are cheaper and were beneficial to them after resource integration. Nevertheless, circumstances changed. With lesser demand, financial difficulties, objection from domestic enterprises and other factors, they were required to adjust the criteria for integration in order to maximize their national benefits. At that time, we lacked such a vision. We failed to understand the concept of maximizing national benefits, and viewed such integration as an international division of labor. Unprepared for change, especially for sudden change, we were reduced to a passive position. We must recognize the need for a more strategic vision instead of just blaming nations with protectionist policies.

Since the policy of reform and opening-up, we have suffered from such dilemmas. According to one of the central leaders, we should both open our door to the outside world and develop the national economy. This is harder to effect than it sounds. Since the policy of reform and opening-up, various industries in China have developed in different courses with different strategies. Their stories have taught us lessons.

We could make a comparison between the household appliances industry and the automobile industry. Although they appear well-developed, what is at their core is rather different. The Chinese self-owned brands of household appliances are not only playing a leading role in the domestic market, but also taking increasing shares in the international market, so the household appliances industry is fairly healthy. The automobile industry of China, however, faces great obstacles. The self-owned brands of China are struggling, because foreign brands have almost monopolized the market for medium- and high-grade automobiles. Currently, though the annual sales of automobiles has reached 20 million in China, the self-owned brands only account for 2% of the total sale of medium- and high-grade automobiles, even with displacement at and over 2.0 liters and good performance. The seemingly booming automobile market in China is actually a feast for foreign enterprises.

In the past, we attempted to sacrifice the market in exchange for technologies, which seemed reasonable at the time. Consequently, the market is opened to the outside world while we still fail to acquire the technologies we want. Why? It is because of a lack of strategic vision; we do not understand the nature of capital, technologies and brands.

It is not an exaggeration to describe technology as the lifeblood of industry. Therefore, although large enterprises in developed nations expect to occupy the Chinese market, they never abandon technology for the sake of market share. According to the Wild Goose Strategy proposed by Kaname Akamatsu (1897–1974), a Japanese economist, Japan is the leading wild goose, while China is the ending wild goose. In order to make China give up independent R&D and intellectual property, Japan sells its outdated technology to China at low prices, or even gives it away for free. The 1990s witnessed the adjustment of international industries. Since then, in order to reduce costs, multinational corporations transferred the processing and assembling industries, which were of low added-value, to developing nations, and China was the first choice for them. At that time, some economists used the "free trade theory" to explain the trend of China's economy. They believed that considering comparative advantages, China should develop itself into a "world factory." As a result, a large number of Chinese enterprises gave up their efforts to create their own IPRs, busy doing assembly and processing for multinationals.

Capital brings power. It is clear that foreign enterprises hold as many shares as they can when they establish joint-venture enterprises in China. Theoretically, it is we who make use of foreign capital; but practically, it is the foreign capital that makes use of domestic capital, because China's capital follows the orders of the foreign capital.

Brands bring market. By eliminating Chinese brands, foreign companies make their brands prominent and so dominate the market on a permanent basis. For example, consumers in China have been captivated by foreign automobile brands. With automobiles of the same quality produced by foreign and domestic brands, consumers will buy domestic vehicles only if the price is 30% lower than foreign ones. How can Chinese brands develop well in such a situation?

The Chinese PV industry should by no means repeat the mistakes of the automobile industry. When it adjusted its PV policies in 2012, the Chinese government sent a clear and strong signal to the world: that the domestic PV industry of China would develop by leaps and bounds, and that the Chinese government would work out favorable policies for the PV industry. This was certainly good news for Chinese PV enterprises, and also for foreign PV enterprises. At such a pivotal moment, we have to determine our strategies. In other words, in what way will China's PV industry participate in global competition?

> Capital, technologies and brands are the most critical factors in global resource integration. Chinese PV enterprises, all industries, really, can participate in global competition in a brand-new way if they understand and acquire these factors.

Two M&A cases in the PV industry are worthy of attention, one involving a German enterprise acquiring a Chinese enterprise, and the other a Chinese enterprise's acquisition of a German one.

On December 21st, 2012, SMA Solar Technology Ag (henceforth referred to as SMA) acquired 72.5% of Jiangsu Zeversolar New Energy Co., Ltd.'s (henceforth referred to as Zeversolar) shares. SMA is a global giant in manufacturing inverters. It used to occupy 70% of the inverter market in its heyday; even now it still occupies 40%. Zeversolar is one of the top 5 Chinese inverter producers.

PV inverters are essential equipment in solar power stations. Direct current generated by solar cell modules is converted to alternating current for integration into the grid. China's PV industry produces a large number of cell modules. Its production of inverters only accounts for 10% of the global total.

It is widely believed by insiders that this acquisition symbolizes a long-term strategy; its timing was also right.

It is little known that SolarWorld, the German company which initiated the investigations of Chinese PV products, is a good partner of SMA. In a sense, the dilemma caused by the investigations of Chinese PV enterprises helped SMA acquire the shares of Zeversolar at a low price. The fair value of Zeversolar as estimated by experts should be about 600 million yuan. However, its fair value was lowered to 220 million when acquired. In other words, SMA spent only 160 million yuan to acquire 72.5% of Zeversolar's shares, a 70% discount.

Why is SMA's acquisition considered as a long-term strategy? In my opinion, SMA accurately predicted that China's PV industry would develop well, and aimed at the inverter, which China is not good at manufacturing. Entering the Chinese market through acquisition of a Chinese enterprise, SMA was not only able to grasp on-coming market opportunities, but also gain considerable subsidies from the Chinese government.

We cannot blame foreign enterprises for their strategic insight, as we cannot blame rivals for their strength in athletic contests. What we can do is improve ourselves by learning from our rivals.

The other M&A case, Hanergy's acquisition of MiaSole, an American thin-film solar enterprise, is different in approach but equally important.

On January 10, 2013, *Securities Daily* published an article entitled *Domestic Enterprises Massively Acquire Foreign Enterprises for Their Advanced Technology*. A day earlier, Hanergy Holding Group Ltd. had officially announced in Beijing the complete acquisition of MiaSole, a leading producer in CIGS thin-film solar modules. At the time, the MP conversion efficiency of its thin-film PV modules had already reached 15.5%, and it was expected to reach 17% and more in 2014. In addition, production cost should be reduced to USD 0.5/kWh. In this way, Hanergy obtained the most effective CIGS technologies for conversion in the world. Hanergy became a leader in both scale and technology.

If SMA's acquisition of Zeversolar was based on the strategic prediction that the Chinese PV market would develop significantly, then Hanergy's acquisition of Miasole shows a strategic will to keep a foothold in the domestic market to integrate global resources, as well as to depend on self-owned brands to integrate more advanced technologies.

From Putting People First to the Harmony Between Man and Nature

The PV industry, an emerging industry, evolves quickly with each day. In order to predict its future trend and formulate strategies for its development, we are required to keep pace with the times and even to take the lead. This means that many outdated concepts will be updated, many narrow ones widened, and many wrong ones changed.

Fundamentally speaking, to promote the new energy revolution is a critical move for putting the *Scientific Outlook on Development* into practice. This *Outlook* advocates putting people first in an all-encompassing, coordinated and sustainable development. In other words, it's basis is the primacy of people and the harmony between man and nature.

Putting people first is advocated in order to correct the concept of putting money first. In order to make society prosperous and the nation powerful, we certainly have to develop the economy, to pursue profits, and to generate wealth. However, this wealth should help improve people's standard of living. If the wealth we generate is not helpful for achieving that purpose, and even does harm, we've embarked on the wrong road. Such economic activities cannot be defined as positive.

Take urban construction as an example. In some cities, buildings and factories are built randomly without proper planning. As a result, this irrational layout brings much inconvenience into peoples' lives. The GDPs of these cities do increase, but people struggle, because this development mode fails to put people first. Some enterprises emit severe waste water at will. They may earn profits and the local GDPs may increase, but the environment is damaged and people's health is badly affected. This type of development is bad also. Similar examples can be found everywhere. Putting people first should be the primary principle for us to follow in our economy, in its development and in a harmonious society.

We bring forward the harmony between man and nature, because we focus not just on people. In the narrow view of putting people first, all human activities should be judged by their benefit to human beings but the ability of the planet to support other species is ignored. Nowadays, slogans such as "man defeating nature," "declaring war on nature," "conquering nature," which were popular in the past, are already outmoded. When the human population was small and our force was weak, these slogans helped our confidence: we had to take the necessities of life from nature. But now the world's population has reached 7 billion and industry is global and modern. However, a large number of plants and animals have become extinct, and environmental pollution and global warming are quite serious issues. Nature has warned us and begun to punish us. The idea of caring only about people must be changed. Many years ago, *The Book of Changes* highlighted the notion of harmony between man and nature, a scientific concept which is still attractive. Without that harmony, we cannot achieve sustainable development. Only by establishing the rule of development in a scientific manner, putting people first but reaching a match of man with nature, can we truly understand the significance of the new energy revolution and the PV revolution.

Replacing Conventional Thought with Innovative Ideas

The natural properties of solar power provide the basis for this harmony. We have to first clear up some misunderstandings when we attempt to turn it into a commercial product accessible to everyone. Conventionally, solar power is not the first energy form coming to mind.

In the conventional view, we reflect on problems and seek solutions based on observation of the past. Innovative thinking helps us break through our old patterns of thought and find a new way.

> Today we have to tackle energy problems with new ideas. Making the development of the PV industry a national strategy is one. It is not only a new way out of the energy trap for China and other countries around the world, but also a firm foothold for China to use to rise as a power.

When referring to the differences between Japanese and Americans, people often say that the Japanese are good at learning, while Americans are good at innovating. In the 1970s, Japanese firms acquired the best practices in some of the key American industries and even surpassed the US in many areas. In addition, the Japanese bought many assets from American companies; for example, Disney. To offset the loss of conventional advantages, Americans chose innovation. Thanks to their efforts, they made breakthroughs in IT application. They developed the PC from larger computers and kickstarted the Internet and then the whole IT industry. We can see that innovation helps the United States to maintain its status as a superpower in the world.

If we think about energy needs in the conventional way, we will look for more of the same forms in the same ways. We will dig for more coal if we need it. We will explore for more oil if the demand is there. We will mine on the sea floor if land resources are depleted. We will import more and larger amounts of oil or acquire more foreign oil fields if our own supply cannot meet demand. We will transmit natural gas and electric power from western China to the east because western China is rich in natural resources while the east is largely an energy consumer. We will advocate for energy conservation and emissions reduction because of the severe pollution caused by fossil fuels.

We cannot say these ways of approaching the issue are wrong; they have helped us overcome difficulties in the past. However, they cannot root out the problem, for the inherent defects of fossil fuels are unchangeable. We cannot resolve the conflicts between consumption and reserves of energy, between consumption and distribution, and between consumption and environmental protection if we are trapped in an outdated thought pattern.

Nevertheless, we can break the existing energy structure and solve this dilemma if we seek solutions in a new manner. Since the defects of fossil fuels cannot be overcome, we can seek and take advantage of new energies and gradually replace conventional energy sources with new ones.

Aiming at Surpassing Instead of Catching Up with Others

Currently, the PV industry in China is far from mature, so we still need to learn from the US and Europe, particularly in how to build up the market. We are confident nevertheless that China will take a leading position in the new energy revolution. It will play a leading role in the world as long as we are determined to surpass others.

Throughout world history, followers have surpassed the forerunners sooner or later. Of course, historical reasons and conditions are needed before that comes to pass, but there are precedents that we can follow. The followers have been able to make use of the fruits generated by the forerunners, with far less effort. In addition, lessons from the leading countries are always valuable for those lagging behind. With this in mind, we can find a new way.

For latecomers, one of the key steps is to set up strategies based on taking advantage of their very backwardness.

The strategy consists of two steps, catching up and surpassing. We will fall further behind if we do not try to narrow the gap. But if we stop after catching up without pressing on, we will remain followers for a longer period. It is our ultimate aim to surpass the forerunners.

In reality, the surpassing strategy is sometimes misunderstood as simply chasing after the forerunners. Countries do move forward in this way compared with where they were in the past. However, they can stay backward, perhaps forever, when compared with more developed countries.

We have to be ambitious to take the lead. Even if it is hard to do so in every field in a short period of time, it can be done in some areas. Then we gradually apply the lessons generally when we are more experienced.

In fact, China has outrun other nations over the last three decades. For instance, the output of more than 170 Chinese products is the largest in the world. More output leads to bigger economic aggregate. The economic aggregate of China is now bigger than that of Russia, Canada, France, Britain, Germany, and Japan. Currently people are trying to predict when China will take the place of the United States as the world's largest economy.

Some think it's mission impossible, because they do not understand the time-honored philosophy that everyone has his advantages. Though we cannot exceed our rivals in each indicator, we may have a bigger economy if we consider the aggregate.

Today, China stays inferior to others in technologies in many fields, but superior to others in manufacturing. Technologies and manufacturing are connected. Technologies not put into manufacturing are useless. With its large manufacturing scale, it is practical for Chinese enterprises to aim for superior technologies. Actually, some Chinese enterprises have acquired world-class technologies and innovate at the same time. These businesses put them into large-scale manufacturing and grasp them very quickly.

The high-speed railway is a good example where China has astonishingly surpassed developed nations technically. Sany Heavy Industry Co., Ltd. is a latecomer who has moved faster in the area of construction machinery. Currently the concrete conveyors it produces can eject material faster than those of any other manufacturer in the world. In addition, this company showed its strength to the whole world during the rescue following the Fukushima nuclear disaster in Japan in 2011.

In my opinion, Chinese PV enterprises will be leaders in the global industrial chain sooner or later. This is very possible in the future. We can surpass others as long as we have the courage and willingness to do so.

What Government Did, Did Not Do, and Did Too Much Of

Since 2001, Suntech has experienced surprising ups and downs. The company was born and expanded with the help of the government and was reshuffled mainly under the leadership of the government when it was on the verge of bankruptcy. That is an accurate description of the relationship between companies and local governments.

When we think about the experiences and lessons of China's PV industry in 2012, the role of local government has attracted the most attention. At this stage, it is imperative for us to answer the following question: what is the proper role for local government to play?

We hope that the visible hand can be more effective. To this end, government should put in place strategies, plans, policies and measures.

What Governments Did Too Much Of

In March 2013, Suntech, a flagship in China's PV industry, went bankrupt. This will forever affect China's PV industry; and, perhaps, it is a new starting point for growth.

After 2010, the production capacity of the PV industry expanded extraordinarily, and leading companies enlarged their capacities actively to grab more market share. Furthermore, local governments also played a key role. Take Suntech for example: established in 2001, it was listed on the New York Stock Exchange in 2005 and earned its founder the title of richest person in China. Yet, until March 2013, when it declared bankruptcy because of insolvency, the municipal government of Wuxi had always supported Suntech, a "business card" in the local community.

In 2000, Mr. Shi Zhongrong, boss of Suntech, returned to China to start his business. He traveled to several cities, meeting the people who were in charge there. He told them that his own products could be very profitable, saying: "Give me eight million dollars, I will give you the largest enterprise in the world." However, none of them dared accept such an offer, as at that time they had never heard of solar energy as an industry.

Fortunately, the government of Wuxi City was very optimistic about his project. In September 2001, under the leadership of the local government, eight large state-owned enterprises (SOEs) jointly financed USD 6 million and became major shareholders in the new company, of which Mr. Shi bought 25% of shares in tech stock and USD 400,000 in cash. Suntech Solar Power Energy Company (Suntech Power) came into being.[10] Before that, the local government's largest involvement in entrepreneurship programs was no more than 8 million yuan. This is the starting point of Mr. Shi's journey to fortune but also a signal event in the history of the Chinese PV industry.

In December 2005, Suntech, successfully listed on the New York Stock Exchange, became a symbol of Wuxi. In 2010, its production capacity of solar cell pieces expanded to 1.8 GW. Even after it became the largest solar panel manufacturer in the world, the local government was still not satisfied with its size.

Therefore, at the beginning of 2011, in order for Suntech to continue its expansion, the government put forward a plan for doubling capacity within five years and it earmarked dozens of acres of land for the company. In turn, the company was required to build another factory employing 50,000 workers before the deadline. The Wuxi New District also invested with the company in the construction of an industrial base for PV, which was expected to sell PV products worth 10 billion yuan in 2012. However, in 2010, the company generated no more than USD 2.9 billion, or about 20 billion yuan, from its main business.

Suntech produced 2.4 GW worth of solar cell pieces in 2011, one-third more than the 1.8 GW of 2010. In 2010, its fixed assets were expanded from USD 777.6 million in 2009 to USD 1.3262 billion, increasing by as much as 70.55%. In 2011, this number further rose to USD 1.5692 billion. How to finance such an investment? The company borrowed from banks, of course. In 2009, Suntech borrowed USD 938.4 million from banks. That amount increased to USD 1.5641 billion in 2010 and USD 1.7067 billion in 2011. Most of these were short-term loans whose worth increased to USD 1.5734 billion by the end of 2011, about 92.2% of the USD 1.7067 billion total. Why did banks have so much confidence in the company? Besides an optimistic attitude towards the company itself, it was the municipal government's credit enforcement or, in some cases, that of the provincial government that really mattered.

In 2011, after the expansion, Suntech produced over 2GW of solar cell pieces per year, a one-third increase in comparison with 1.5 GW in 2010. Unfortunately, the price of PV products kept diving in that year, causing the company's sales revenue to increase by only 8.4%, or USD 3.147 billion. Also in 2010, the company's net loss totaled USD 1.018 billion, but in 2010, its profit totaled USD 238 million for the first time.

In 2012, the company's output slightly reduced, but still stood at 1.8 GW. Its operating revenue nonetheless decreased to USD 1.625 billion, a 48% drop. The huge loss in 2011 and the continuing deterioration of business in 2012 finally exposed the financial risks which the company had accumulated during its fast expansion. By the end of March 2012, the company's total assets were worth USD 4.3786 billion, but its debt amounted to USD 3.5754 billion, or about 81.66% of total assets. The company's net assets dropped to USD 803.2 million.

Even worse, the cash and cash equivalents in its hands were only worth USD 473.7 million. In the total debt, short-term liability was as high as USD 1.5746 billion. There was also another USD 557.8 million of convertible bonds, which were overdue on March 15th. As a result, Suntech came to the edge of the capital chain break.

In September 2012, Suntech began to lay off workers, cutting the production of solar cell pieces to 1.8 GW from 2011's 2.4 GW. At about the same time, it was forced by its debtor banks to pay its debts. All of a sudden, the rumor of its bankruptcy spread quickly.

In such an emergency, the local governments intervened again. At the end of September 2012, the leadership of the city arrived at Suntech, starting a comprehensive support campaign. With a push by the municipal government, the local branch of the Bank of China extended another 200 million yuan in loans, which injected energy into the company in a time of huge financial difficulties. However, this was only the tip of the iceberg. In March 2013, the Wuxi Suntech Solar Power Company (Wuxi Suntech in short), the core company of Suntech Power, finally declared bankruptcy and started recombination.

Earlier, in the second half of 2012, Wuxi Suntech had already faced difficulties when the State Assets Supervision and Administration Commission of the city intervened through Wuxi Guolian Development Group Corp. Ltd., a company solely funded by the central government. At the time, it was believed that this company would succeed Wuxi Suntech.

However, in August of 2013, Mr. Huang Qing, member of the Party Committee of Wuxi City and Deputy Mayor said in an interview, "The Wuxi government is looking for the most powerful strategic investors to join the restructuring to fully optimize bankruptcy of Wuxi Suntech.... Now, five companies have been contacted, including three private and two state-owned ones. In the future, Wuxi Suntech will not be cleared."[11]

Although the restructuring of the company did not settle down by August of 2013, it let people see a relatively clear direction for the market-based restructuring. Perhaps after that, the company and local government would form a healthier relationship. It was not bad for the company which has finished its primitive accumulation.

Let Government Play Its Role

The boom and bust of Suntech and LDK Solar are not the first such cases. They just expose in a radical manner the typical characteristics of China's PV companies in the period of "brutal growth." How can we judge the role of government, particularly, local governments, in this process? This is no yes or no question. To answer, we need to return to our starting point, namely, the relationship between government and the market.

As we have mentioned previously, the relationship between government and the market should be as follows: the market should play its decisive role alone, while the government should supplement the ignored or less developed fields in the market through macroeconomic control. However, once the government intervenes in economic activities, it is fair to wonder how actively it should be involved. We are left to ponder what governments did, did not do, and did too much of. Before 2011, it was generally believed that local governments in China were very involved in developing the PV industry. They even vied with each other to plan the development of this industry in their region, and spared no effort in supporting the growth of local PV companies. In contrast, the central government was more cautious.

Some people believe that overcapacity in the PV industry was due to some local governments overplaying their roles. Due to governments' in-depth intervention and overactive support, developers rushed into this sector and expanded their production capacity sharply, which ultimately led to the final crisis we have witnessed. Others believe that the government didn't do enough. For example, the government didn't provide proper planning and guidance for healthy growth. The raw material suppliers and purchasers of China's PV products are mainly in foreign countries; this creates potential risks. The government should have given timely guidance and made adjustments accordingly.

These ideas seem to be in conflict with each other, but they all make sense. They both argue that the government didn't play a proper role. It either did not do enough or did too much.

The correct role of government is critical to the development of this industry. In microeconomics, companies should be managed independently, and take responsibility for their profits and loss. The government should never put the cart before the horse, or exert administrative intervention by force. However, in an industry, there are many enterprises, and the relationship between this industry and other industries should be coordinated. Its position and the role it should play in the whole national economy should be considered. All these are macro and mid-level problems which cannot be solved solely by companies themselves. Instead, the government should have its part in control.

The PV industry has many unique features, such as the asymmetry of the interest relationship. It is also fundamental, comprehensive, and related to other industries. Therefore, it is impossible to deal with that many problems by allowing this industry to develop solely in relation to other enterprises or the market. If the national government issues an energy strategy relying on traditional fossil fuel instead of new energy, how can these new energies gain a favorable position? Depending solely on market competition, can PV startups win the battle against the strong companies in the traditional fuel sector? Without proper policies to support this industry, and by relying solely on price competition will the PV market prosper in the future? Such questions remain unanswerable, at least in the relative long-term.

We should realize that several years ago, China's PV industry developed mainly because of the spill-over effect from the new energy revolution promoted by foreign governments, and European governments in particular. It was not the result of pure market competition. In order to change their energy mix, European governments, particularly the German government, strived to develop the new energy system focusing on PV. By giving high subsidies to new energy developers, the market demand for PV product was released. The PV products from China were very competitive in European markets because they were cheap and manufacturing in China is inexpensive. Of course, the premise for this was the preferential policies from the local governments for new energy growth.

Let's have a look at the domestic market here. Since local governments didn't do what they should have done before 2012, the national government didn't have a proper, clear strategy and policies on the PV industry. If there were any such policies, we can now affirm that they were not very useful. As a result, the domestic market was not well developed. Even worse, the anti-dumping and anti-subsidy investigations from foreign countries have forced the Chinese PV industry to the brink of extinction.

Therefore, by looking at the roles played by foreign governments and the Chinese government, we can see huge and even critical influence on the growth of this industry.

To further understand and tolerate the governments' actions, we may analyze more deeply their role in developing the PV industry. It is understandable that local governments try their best to boost the PV industry so as to develop the local economy. An active government role is needed for this industry's future prosperity. It's also understandable that the central government failed to realize that it needs to play a bigger role in the early life of this industry. Generally speaking, it takes time to understand an energy revolution and its industry. Starting to realize the prospects for this industry, the Chinese government has instituted a lot of supportive policies. For example, in 2012, a series of pro-PV policies were released.

Based on the above analysis, we come to the conclusion that as an emerging industry closely related to the country's comprehensive economy, China's PV industry should see the government play a proper role so that it's growth will be smooth. The market's invisible hand should play a bigger role in this process.

Letting Strategies Play Their Due Role

To allow the invisible hand to play its role correctly, the government should put in place the correct strategies, plans, policies and measures.

Strategies show directions and tell us what to do and why. The national government should research this industry in-depth, and subsequently come up with more insightful and clearer strategy on new energies. These strategies should be holistic strategies, in touch with the healthy and sustained growth of the national economy as well as social progress.

The preconditions for such a strategy include predicting the trends that the world will experience in the third Industrial Revolution, comprehensively understanding the global energy structure, and recognizing the disadvantages that fossil fuel energy is unable to overcome. Moreover, looking at the energy bottlenecks in the way of China's sustained economic growth, the government should understand the PV industry in a comprehensive manner, and realize the importance of developing this industry in different aspects, in the short-term as well as the long-term.

This strategy should not only aim at increasing the quantity of new energies, but also at changing the entire energy structure and system. This strategy should aim at an energy revolution: that is, the PV revolution.

Based on this strategy, the slogan for the new energy revolution should be: "changing the world with new energies." As for the strategy for new energy revolution, it can be called the *strategy of energy substitution*. That is to say, the reason why we want to develop new energies is to replace fossil fuel and change the world accordingly.

Here, the keyword is *substitution*. To correctly understand it, we need to know what will replace what, in what manner, and how to judge the operation's success.

First of all, what will replace what? Obviously, it is for new energies to take the place of traditional fossil fuels, instead of replacing one kind of fossil fuel with another. Due to a misunderstanding, some people regard some of the newly discovered energies or energies attracting recent attention, like shale gases, as new energy.

When will we be able to affirm that traditional fossil fuel has been replaced by new energies? Simply put, by improving the proportion of new energies in the energy structure, not by simply avoiding the use of fossil fuel, new energies will play a dominant role.

However, in the current energy structure, fossil fuels take the lion's share while new energies are still supplementary or auxiliary ones. The energy revolution must change this structure. In replacing fossil fuels, new energies will gain more importance, and finally take the place they should.

Currently, many countries regard maintaining the dominance of fossil fuel as the most important task for the future. It's easy to demonstrate this by comparing the input countries make in looking for oil and their investment in developing new energies. Dispute over oil has never cooled down. Even worse, shale gas, as a new kind of fossil fuel, is hotly discussed around the world. One of the consequences is less input from the American government in new energy.

At this stage, we don't and shouldn't oppose efforts to maintain the fossil fuel supply. What, then, should be the core of our strategy? Should we aim at maintaining the dominance of fossil fuels or at developing new sources?

Other countries actually came up with a substitution strategy many years ago. As we have mentioned, President Obama once predicted that by 2025, renewable energies in the US would make up 25% of the total consumption of energy in the whole world. UN Secretary General Ban Ki-moon also predicted that by 2030, renewable energies around the world would make up 30% of total energy consumption. And I believe that by 2035, clean energy will account for 50% of the total consumption of primary energy in the world. This is not an illusion, but an objective we can achieve through our efforts.

Of course, an objective in itself is not enough. In what manner will new energies replace their predecessors? We need to work out a new substitution strategy which correlates with reality.

We must be quite aware that new energies will not take the reins in one day, but rather that such a substitution will occur over a long period of time. As a result, over such a long period, the absolute volume of both fossil fuel and new energies will keep growing.

However, in this process, the relative weights of new and fossil energies will continuously change. As the share of new energies increases faster than fossil fuels, the former will gradually gain more and more, before replacing the latter as the dominant player. After that point, the absolute volume of fossil fuel we consume will begin to drop.

Such a substitution generally can be divided into three stages. The first stage will be substitution in speed. That means the absolute quantity of fossil fuels and new energies will increase at the same time, but the former faster than the latter. The second stage will be a substitution in volume. The absolute volume increase of new energies will be larger than that of fossil fuels. The third stage will be one in which the dominant place will be obtained. The share of new energies will be larger than that of fossil fuel. Moreover, the volume of fossil fuel consumed will gradually decrease and make room for new energies.

In the process of moving from a quantitative change to a qualitative change, we must understand the following key points.

First, the growth rate of new energies is key. In the future, fossil fuel will be replaced not because of a reduction in consumption but rather because consumption of new energies is increased. The growth rate of new energies will be very important indeed. To make this happen faster, we need to strike a balance. On one hand, we should spare no efforts in building surface stations, and on the other hand we should build distributed stations. In the future energy mix, solar power will be used mainly through distributed stations. In the initial stage, however, large-scale and concentrated PV generation should be the vanguard and the main force. This is due to government support: in such a centralized manner, a station can be finished within half a year, allowing companies to enjoy easy operation and ready benefit.

Second, the replacement should happen first in advantageous regions. To encourage the use of new energies, China should establish them in regions with the biggest advantages. For example, the province of Qinghai, an experimental base selected by the State Council, witnessed a fast increase in PV stations. In 2011, the province increased its connected PV capacity by 1.003 GW, about 50% of the total installed capacity in the country. And this number more than doubled the total newly installed capacity in 2010 (0.53GW). Similarly, 10 other provinces and autonomous regions, including Ningxia, Jingsu, Gansu, Xinjiang, Shandong, Inner Mongolia, Hebei, Tibet and Shanxi, have also attached equal importance to PV development. The above mentioned regions will see faster growth in PV.

Third, the replacement process should start with fossil fuels, which China lacks most. We all know that among fossil fuels, what China really needs most is oil. Moreover, with car ownership increasing rapidly year over year, the oil shortage has become all the more obvious. This gap has to be filled by new energies. China has begun to develop electric cars and related industries on a large scale. Yet it hasn't broken down the old mix of traditional energies, because the electricity used is generated from burning coal—this is just the replacement of one fossil fuel by another. Only by making the link between electric vehicles and new energies can we fundamentally make a breakthrough.

Fourth, more big players should be enlisted, particularly users of fossil fuel, to replace fossil fuel by developing new energies independently. For that reason, I believe that distributed PV stations should play a major role in the replacement strategy. We need to mobilize a variety of players, including governmental organizations, enterprises, factories, citizens, the military as well as public facilities. That is, those with the ability to build power generation facilities independently should be encouraged and supported to join the energy revolution.

However, it should be emphasized that the basic strategy in the process of replacement should be a strategy of common interests, because the energy revolution has no enemies. It will be an industrial revolution, benefiting all of society. It will be a common zone of interest for states, enterprises and ordinary people themselves.

From the industry's point of view, we may think new energy companies and companies developing fossil fuels are meant to be enemies. This is wrong. In fact, in this revolution, they will be partners, because the companies themselves are not replaced, fossil fuel is. Even then, the objective is not to replace fossil fuel per se, but to change consumption and utilization patterns. Fossil fuels are a valuable resource of nature, and they should be used in more valuable ways. During the replacement, fossil fuel companies will face less supply pressure and make more profit by using the fuel remaining as chemical raw materials. Through the energy revolution, power companies who provide secondary energy will be able to depend on renewable energies to improve their operational efficiency based on more efficient and safer supply models. More importantly, with energy substitution, the national economy can develop in a sustainable manner, which is the basis for company growth. Therefore, the energy revolution does not aim to create rivals, but partners, who together work towards upgrading industries and accelerating economic growth.

By understanding that such a revolution is of common interest, those involved can be more coordinated, play their own role, and make full use of their advantages and resources in strategic substitution. In this way, they can join forces and get the result expected.

Of course, before clarifying our strategies, we must make clear plans. Without plans, strategies cannot be thoroughly implemented. Plans help us set up targets, and determine the basic means and major steps for each implementation phase. They provide guidance and reference points for more effective coordination on all sides.

The 12th Five-Year Plan has started planning for new energies and the PV industry. However, we need to implement a new understanding of PV in relation to new situations and revise the plan accordingly. More importantly, PV shouldn't be regarded as an auxiliary energy anymore, but as an alternative energy source. We need to rethink the relationship between fossil fuel and new energies, and we need to put more attention and input into the PV industry for faster growth.

Letting Policies Play Their Role

Strategies and plans specify the targets, road maps, means and steps. If you want to turn these into reality, however, you also need driving policies.

As we have mentioned before, putting in place the right policies means establishing a policy system with internal connections. However, there are some challenges. First, how should we make use of the three policy leverages (financing, taxation and pricing)? Financially speaking, we need to know how much, where, and for how long we should finance. We should know how much, how long and for whom we should reduce taxes. As prices are closely related to financial and taxation policies, we should decide whether it is up to the national government or individual regions to define prices. How should we let the market define prices in the place of government? All these challenges need to be faced and solved.

Second, how should we define the focus of policy? For the PV industry, favorable policies are the supportive and preferential ones. These policies have three focal points: production, supported by tax exemption, tax reduction, and subsidies of soft loans for PV producers; construction, in which builders of and investors in (including individuals) PV stations will be given financial subsidies; utilization, in which proper on-grid tariffs should be defined to increase return on investment and mobilize investors.

Different focal points for these preferential policies will have different impact. If we only stimulate production instead of construction, capacity surplus will be more obvious. On the contrary, if we only promote construction without enough consumption, substitution cannot be realized. If we only encourage the use of PV and other new energies, it will be very difficult for this industry to take off because it will take a long time to generate benefit. The government needs to consider where it should put its emphasis or how to balance the emphasis. The government should also clarify the intensity, timing, differences among, beneficial regions and targets of these policies.

At present, the large scale application of PV in China should be the key point, as well as the focus of these policies.

> Now the PV industry needs policy support from the central government, but the purpose of such subsidies is so that it won't need subsidy in the future. If the industry survives by depending on subsidies, the market has no future. By subsidizing, we hope to promote the progress of the industry and technology, and realize the parity of prices on grid, so that citizens can afford the electricity generated by solar power.

Third, how should we address the relationship between generation and grid integration? By far the most important condition for the fast growth of solar power is its smooth integration into the grid. Large-scale solar stations and, more importantly, small ones in tens of thousands of households need to be connected to the grid. Otherwise our plans won't be realized.

When we say smooth integration, we mean not only the smooth connection between PV stations and the power grid, but also their coordinated operation.

For this purpose, both stations and grid need to get used to each other in a dynamic connection. Some problems remain which need to be solved technically and operationally, therefore policy leverage is needed to coordinate interests on all sides. For example, if the on-grid tariff of PV is too low, there will be no incentive for use and no possibility for this industry to grow. If the price is too high, the power grid may suffer losses. The more they buy from PV stations, the more loss they will have. As a result, enthusiasm for the grid will be dampened.

Fourth, how should we extend the reform of the electricity system? Grid parity refers to the relationship between the price of PV and existing energies. On one hand, the cost of PV generation must be reduced so as to reduce PV electricity prices to the level of current energy prices. On the other hand, we should check whether the pricing of existing energies is proper or not. If it is too low because of policies, which makes it difficult to reach parity with PV prices, then PV generation should not take all the blame.

In order to cut production costs for companies as well as to accommodate citizens, the Chinese government has kept tariffs low. Moreover, because China depends on thermal power generation, electricity prices are directly linked to those of coal. To cut tariffs, we need to lower the price of coal. Globally speaking, the ratio of the prices of the major energy forms, coal, oil, natural gas, and electricity, should be 51:67:77:100, but the current Chinese ratio is 7:49:37:100. Obviously, coal prices in China are extremely low. However, if coal prices increase and the tariff remains unchanged, this will result in conflicts between generators and coal suppliers. For a long time in the past, only administrative forces could define coal prices, in order to keep thermal plants profitable. Under these circumstances, it is unreasonable to lower the market price of PV electricity to the level of coal-fired electricity. It is a problem which cannot be solved immediately, but it should be taken into consideration when talking about parity. Only by doing so can the perspectives of government and society come closer to reality on PV.

In fact, to solve the problem of coal electricity, the national government has begun to implement the step tariff. That is, if users go over a certain level of electricity consumption, their tariff will be increased. This reform has not been totally put in place yet.

At the same time, we need to realize that to push the energy revolution, as well as the large-scale integration of PV and other forms of new energies into the grid, we must develop distributed stations in particular. Only thus can we trigger revolutionary changes in the grid. China's electrical power system has made big steps in its reform, but there still are plenty of problems and challenges waiting to be solved. In future reform efforts, the grid should consider the full extent of this revolution, and evolve into a smart grid.

Policy on new energies is not only meant to support the development of the PV industry, but also to affect the whole picture of energy and the economy in China. Policy should be made and implemented by related functional government departments, in collaboration with other departments. Through field visits and brainstorming, these departments need to study, stipulate and implement their policies. Therefore, relevant ministries and even the State Council should take the lead by establishing a PV leading group to solve the strategic challenges of energy substitution.

Reform is going to solve problems as they emerge in the process of growth. With favorable policies, reform can be a booster for growth. In a sense, this is an area which will test the ability of government to lead the revolution. It is a task which can only be undertaken by the government; it is a chance to demonstrate leadership.

Of course, after making strategies, plans and policies, concrete measures should be taken to implement them. There are overarching measures, partial measures and individual measures. Overarching measures are meant to promote comprehensive development, while partial and individual measures target regional problems or individual challenges.

In the solar power sector, the Chinese government has taken two important measures: the "Golden Sun Demonstration Project" and the Building Integrated PV (BIPV) project. The Chinese government has earmarked over 10 billion yuan of financial subsidies for these measures.

At present, the government is discussing what role the financial industry should play in supporting the PV industry.

The thin-film solar power industry in China boasts unique scale and technical advantages. However, because financial institutions blindly have faith in the polycrystalline silicon capacity surplus and fail to distinguish polysilicon and thin-film solar power, they have implemented restrictions on financing. Unfortunately, companies with the core technologies in thin film have often been negatively affected. In July 2013, the State Council issued *Opinions about Promoting the Healthy Development of the PV Industry*. After that, the financing environment took a turn for the better. Nevertheless, since the thin-film industry, a strategic and emerging industry for China's sustainable growth, is still facing bottlenecks, relevant policies and supportive financial policies are badly needed.

First of all, we hope that these policies can support R&D in core technologies and the industrialization of core equipment. For the film industry which will be in the spotlight in the future, financial regulators must work out proactive supportive financial policies and differentiated financial policies. These policies will enable them to support R&D in key thin-film technologies and in the production of related equipment, extend loans for the domestic production of high-end equipment and formulate favorable tax policies. With these things done, the thin-film PV industry can keep its leading place, and ensure healthy growth in China.

Second, we hope a number of key enterprises in this industry can be made stronger and larger with the help of policies. We should select several solar companies which represent the trend in this industry and own their independent IPRs and core technologies. We need to give them special financial, physical and tax support so that they can keep their leading positions in the industry around the world, and make China's voice in the field of clean energy louder. Otherwise, China may miss another unprecedented opportunity in the energy revolution. If the national government can give special support to leading companies to accelerate the R&D of products, the application of solar power products will spread to each and every corner of our lives. This will change peoples' lives. The solar power industry has the potential to become a pillar industry for this country, driving the growth of many other sectors. For example, Hanergy has helped boost 85 industries, including glass, iron and steel, plastics and logistics as well as 1,026 small and medium-sized enterprises, creating 200,000 jobs.

Thirdly, we hope that BIPV and the development of application products can receive greater support. Since the second half of 2012, the National Development and Research Commission (NDRC) and other ministries have released a series of policies to support the domestic PV market with some tangible results. However, Chinese PV power generation accounts for less than 0.1% of total power consumption, far behind the EU and the US. Moreover, China maintains a single model of large-scale development and centralized transmission of power. For example, 90% of our PV electricity comes from large-scale ground stations. The experience of Europe and America has proved that surface power stations should account for no more than 20% of total power generation, while BIPV should be over 80%. Therefore, BIPV and application products should receive greater support from the national government. Such a move will also benefit restrictions on industrial structure and the transformation of the economic development pattern, as well as set another growth point besides that of urbanization, thus effectively boosting the national economy.

Fourth, we hope that the government can make financial resources more available to the private economy. It should institute intensive reforms to further private economy growth. By encouraging and supporting some large private companies with the technologies and competitiveness to join previously monopolized sectors, we can consolidate the national economy and ensure a sustainable growth. This is called the "catfish effect." Furthermore, doing so allows for a broader development platform for private companies, thus injecting new momentum into the national economy.

Successful Stories in Germany

Since 2000, the Chinese government has released a series of policies directed at the healthy development of the PV industry. Beginning in 2012, favorable policies were issued more frequently. How should we judge the effort of the government? If we need to find a benchmark to judge the strategies, planning, policies and measures of the Chinese government against, the best option will be those of Germany.

In February of 2013, the Chinese Ministry of Commerce published an article called *Introduction and Inspiration of the PV Industry Development in Germany.* The article begins by saying that low carbon growth in recent years has become an important trend in the global economy, and renewable energies, including solar power, have become the goal of the all major economies. Europe, with Germany the leaders that they are, is an important engine for the installed PV market. Studying the German case may prove to be very significant for China.

Germany is still the largest PV market in the globe by far for both newly installed capacity and accumulated installed capacity.* In the past decade, Chinese PV companies have developed rapidly. To a large extent, this is due to the opening of the PV application market in Europe, with Germany as the leading country. Let's take Suntech for example. Even when the European market grew at a slower speed, Suntech's income from the European market accounted for about 45% of its total income between the first quarter of 2011 and the first quarter of 2012.

In fact, Germany is not only the largest PV market globally, but was also once the most important PV producer. German PV companies such as Q-Cells, Scheuten Solar, Solar Hybird, Solon, Odersun and the German branches of other famous enterprises in the world lifted the PV manufacturing industry to its peak. Germany was not only the leader in production volume and capacity, but also in technology and brand building. In fact, cutting edge technologies in this industry are still held by German companies. For example, Solibro, which was bought out by Hanergy Holding Group in June 2012, boasted the leading R&D capacity in CIGS thin-film technologies.

---

*According to Hanergy's researh, China became the largest PV market in 2014.

The German PV industry has taken the world lead mainly due to clear and effective support policies from the German government in the past decade.

First, the German policies had clear targets, adjusting themselves at a steady pace. In 2001, Germany came up with subsidies for solar power. In the past decade, the total amount of subsidy has been over 100 billion euros. The idea behind this is to use policy to support the expansion of the solar industry, lower the cost with the scale effect, increase the technological content in products and finally realize high-speed industrialization.

Second, the role of industrial association was given full play, so as to reflect the appeal of the industry. Berleburger Schaumstoffwerk GmbH (BSW), with around 800 member companies, is a bridge between the industry and the government. It aims at transforming solar industry into a permanent pillar of the energy sector. When Germany adjusted its industrial policies towards solar power, this association represented the interests and demands of companies, as well as advocated a step-by-step steady reduction of subsidies. The German Renewable Energies Agency (AEE) has about 100 member companies. It aims at facilitating communication about generating power from renewable energies. It is committed to the safe supply of energy, innovating, increasing employment, strengthening export potential, lowering cost, protecting the environment and saving resources.

Third, public opinion should be guided and people should be encouraged to participate in the growth of solar power. In order to mobilize the general public to join the solar power industry, Germany has lowered the investment threshold. About two-thirds of projects allow investments of less than 500 euros by German citizens. There are even some projects with a threshold as low as 100 euros. In recent years, there have been over 500 participatory solar projects, with a total investment of about 800 million euros. By the first half of 2012, about 80,000 German citizens were promoting such products through shareholding, among them 90% are power generation projects. AEE believes that cooperation in building solar projects helps increase the efficiency and effectiveness of solar power development and usage, boost the local economy, improve people's awareness of environmental protection, and also allows citizens to enjoy the fruits of renewable energy as soon as possible.

Fourth, with exhibitions as a platform, technologies should be developed first. The Munich International Solar PV exhibition and the biannual European solar PV Exhibition held in Hamburg are important events showing trends in the industry. Exhibitions have become an important platform for understanding new dynamics, expanding business, seeing international markets, and showing new technologies and products. The solar industry in Germany attaches importance to technological innovation, and wishes to improve competitiveness by techno-logical advancement. For example, the German Fraunhofer ISE is the largest of its kind in Europe. The polysilicon cells it developed have created many records in conversion ratios. Its thinness allows for economy in the use of polycrystalline silicon during production. Germany is also actively boosting space generation technologies, and taken the lead in solar generation.

Frankly speaking, the PV industry in Germany is now at a very difficult stage. In recent years, the global economic slowdown, the spread of the European debt crisis, and overcapacity in PV module production with dropping prices together made sure that that industry no Germany was spared. In the first half of 2012, more and more companies declared bankruptcy, including Q-Cells, once the larg-est PV cell manufacturer.

In March of 2012, the German government reduced its subsidies, with adverse repercussions on the German and global PV industry. On one hand, because of the intensifying of the European debt crisis, Eurozone countries faced the pressure of larger and larger deficits. On the other hand, that cut can also be thought of as a measure of the healthy growth of the PV industry in Germany. By adjusting the speed and scope of the PV market, Germany controls newly installed capacity at a proper level. The PV industry in Germany generally sup-ports the reduction in subsidies, because they believe that it is necessary to have a steady and gradual plan to cut subsidies.[12]

In the future, because Germany gave up the development of nuclear power, a gap in the electricity market is bound to appear. Germany will spare no effort in developing the new energy sector. In order to make the situation better, Germany is going to encourage bankruptcy recombination, to select the supe-rior and eliminate the inferior among its domestic companies. It will also turn to foreign countries and the new generation of PV technology.

It's worth noting that against the backdrop of reduced subsidies, the German government has turned its focus to supporting thin-film PV. Just as we mentioned in Chapter 2, thin-film PV efficiency has been gradually improved, and its market share in the EU grows larger and larger. Through the new generation of thin-film, Germany is trying to rebuild its leading position in the world. It is a strategy which we often fail to notice.

The Chinese Government Takes Action

Since its beginning, the Chinese PV industry has been haunted by the necessity of importing raw materials and exporting products. Right now, though about half of the PV stations in the world use generation systems manufactured in China, China itself has only a few PV stations.

After China's PV industry has developed itself from the low end, it has to accept the fact that it will soon become the "world factory" as a result of international competition.

The history of the global PV industry has shown us that no country can develop this industry without governmental support, let alone take the global lead. China's PV industry has already acquired the support of local government, but in the end, it will have to gain strategic support at the national level. Fortunately, the Chinese government is taking action. China has pushed forward the construction of distributed power stations and is leading its PV industry into a bright future.

From the Golden Sun Project to Distributed PV Stations

Foreign experiences can serve China well. Since 2012, the Chinese government has scaled up its support for developing its PV industry that may be encountering difficulties. Germany provides a good example of government support for the PV industry.

There are three main ways German policies support the PV industry. First, the government expanded the domestic market, including large-scale ground PV stations and distributed PV stations by providing subsidies. Second, it improved the overall competitiveness of the PV industry by promoting innovation. Third, it gave full play to industrial organizations and let them voice the demands of the industry. Comparing the Chinese experience with that of Germany, we can pick up insights into how the Chinese government can support the industry through policies.

For the Chinese PV industry, the move to expand the domestic market by providing subsidies is no doubt in most need of policy support. Judging from the development of the domestic market in recent years, China's PV industry is moving in the right direction even though it had a difficult start and several setbacks.

In 2009, the Golden Sun project jointly launched by the Ministry of Finance (MOF), the Ministry of Science and Technology (MOST) and the National Energy Administration (NEA) exerted a profound influence on China's PV industry. It was impressive at that time, as the cost of development in the PV industry remained high, and the power grid was not advanced enough to support new energies. The Golden Sun project not only took the lead in the expansion of the domestic market, but also made breakthroughs in grid-connected

PV systems. In theory, licensed PV projects can send power generated by the PV panels to the utility grid.

However, compared to the huge generation capacity of China's PV industry, Golden Sun is still relatively small. Take the year 2012 for example. That year, the installed capacity of Golden Sun was 4.54 GW, a record high. However, this figure is only one-tenth of the total capacity of the Chinese PV industry, which had at that point reached 40 GW.

It took China 4 years to build the Golden Sun project, but after modifying policies several times and investing billions of yuan, it had to face an awkward situation: the investment did not bring about the expected huge output. According to statistics, by the end of 2012, only 40% of the installed capacity of the project was connected to the utility grid. Even worse, other projects failed to connect to the grid due to a variety of reasons. In addition, the practice of providing subsidies before setting up projects was questioned by the stake-holders. Many experts call for a system in which the government subsidizes the prices of the power connected to the grid. In June 2013, the central government finally announced that it would cancel the Golden Sun project.

After this pilot project, large-scale ground PV stations became the most influential in China's PV industry. Although such PV stations are rarely seen built in the deserts in other countries, and although we cannot ascertain whether such construction is scientific and sustainable, they have become the major development pattern in China's PV industry.

A 10MW grid-connected PV generation program was launched in Dunhuang, in the province of Gansu, for which the NEA organized the bidding. It can be regarded as the beginning of the construction of large-scale ground stations in China. When the NDRC released the new pricing policy for on-grid PV power in August 2011, it provided the very incentives that PV companies needed to build this kind of station. According to that policy, power generated by stations approved for construction before July 1, 2011 or those which have been completed before December 31, 2011 will be sold at a price of 1.15 yuan per kWh. Power generated by the stations which have been approved for construction after July 1, 2011 or those which have been approved for construction before July 1st but are still under construction after December 31, 2011 will be sold at the price of 1 yuan per kWh (except in Tibet).

When this policy was issued, the PV market responded quickly and positively, and PV stations emerged like mushrooms after the rain. PV companies in all provinces spared no effort to build large-scale ground stations as quickly as they could. By the end of 2011, the province of Qinghai alone had produced 1.03 GW of on-grid power generated by PV stations, and there was another 1GW of PV power-generating capacity under construction. By then, the total installed capacity of PV power in China had reached 3.6 GW, 4 times of that prior to 2010.

However, with the rapid construction of large-scale PV stations, much of the power generated was wasted. After these stations were completed, many of them were abandoned since the grid infrastructure was not advanced enough to support such rapid growth. For a long time, dozens of PV stations in Golmud, Qinghai had to suspend operations for four days after every two days of generating power, which means that these stations had to take turns generating power and sending it to the utility grid. This has caused a secondary waste of energy. The owners of the PV stations also complained a lot, believing that the State Grid was to blame. However, as pointed out by an expert in power generation, businesses had ignored the special nature of electricity when they rushed to construct stations. This was the reason why so many stations were abandoned. As a special commodity, electricity is difficult to store, and it needs to be transmitted and used as soon as it is generated. When people construct traditional stations, transmission and distribution facilities are built beforehand. However, people obviously ignored this principle when they rushed to build PV stations. When the construction of PV stations becomes greater than that of power transmission and distribution facilities, the grid can no long transmit all the power generated. As a result, many of the stations were abandoned. That is the reason why the State Grid is not to be blamed for the situation.

As the same time, large-scale ground stations kept showing more shortcomings. For example, these stations require a huge area of land, but they can only provide a limited number of jobs, without mentioning the high cost of long-distance transmission. Also, many stations were in arrears with subsidies. As a result, the construction of PV stations gradually slowed down.

Therefore, as we develop the PV industry, we need to be aware that the State Grid is almost full and that the power generated should be transmitted and used immediately. Since PV power has low energy density, is difficult to store for commercial purposes and can only be generated when there is sunlight, it is more rational and scientific to construct distributed stations near the area where the power is used. In fact, the US and many other developed countries in Europe have adopted this pattern of power generation. In this way, users can construct their own PV stations. When there is a power surplus, they can sell it to the market. When there is not enough power, they can buy it from the grid. There are several clear advantages to such a pattern. Full use is made of the power generated, the stations will not be abandoned, the investment in constructing transmission and distribution facilities will be saved, and energy loss during the transmission of power reduced.[13]

After gaining a clearer understanding of the PV industry, the central government began to manage this industry in a "top-to-bottom" manner. In September 2012, when the PV industry was at a crucial point in development, the NEA issued the *12th Five-Year Plan on the Development of Solar Industry*. For the first time, the plan provided that we should develop distributed PV stations. According to the plan, the total installed capacity of China's PV industry would reach 21GW during the 12th Five-Year Plan, of which distributed stations would account for 10 GW, a figure close to that of large-scale PV stations.

When the NEA specified the direction for the future of the PV industry, the State Grid, which had been questioned by the public for acting passively on the integration of renewable energies such as PV power, also changed its attitude. It has publicly pledged that it would fully support the development of distributed PV stations, and has issued the *Advice on Supporting the Grid Connection of Distributed PV Stations*. Since then, China has taken a leap forward on the road to connecting PV power to the grid, and the entire PV industry has been inspired by such a move. In February 2013, the State Grid issued the *Advice on Supporting the Grid Connection of Distributed Power Sources*, thus expanding its support from PV power to all sources of distributed power.

Based on the current situation, the national government has decided to develop both large-scale ground stations and distributed stations. It is worth noting that the NEA has adjusted several times the development target of the 12th Five-Year Plan on developing the PV industry. The policymakers first set the power generation target at 5 GW, and then raised it to 10 GW. In May 2012, they raised it again to 15 GW. In July, the NEA readjusted the target to 21 GW and announced it to the public in September. However, in January 2013, this target was again raised to 35 GW. As the total installed capacity of PV generation in 2012 was only 8.2 GW, this implies a newly installed capacity of around 10 GW per year for the next 3 years.

Shi Dinghuan, Chairman of the Chinese Renewable Energy Association, points out that China raising the target for installed PV generation capacity several times a year proves that the government has attached great importance to the strategic development of the new energy industry.

Problems We Must Face

On July 15th, 2013, when the Chinese PV industry was encountering difficulties, the government issued *Advice on Promoting the Healthy Development of the PV Industry*, bringing enthusiasm and hope to the PV industry in China.

The *Advice* reiterated the strategic position of the PV industry in China. It said: "Developing the PV industry is very important for restructuring the energy mix, pushing forward the revolution in energy production and consumption as well as promoting the development of ecological civilization." The *Advice* also expressed the necessity of "promoting the healthy development of the PV industry." "Due to international demand in the PV industry falling, exports have encountered difficulties and the development of the PV industry has been unbalanced. PV companies were also having trouble in their operations. At the same time, the PV industry in China was faced with challenges such as overcapacity, disordered market competition, overdependence on foreign demand, a small domestic market, weak application and innovation, slow development of key technologies, equipment and materials, lack of financial support and an adequate subsidizing system, weak industrial management, as well as a bad market environment. Therefore, the PV industry was having a hard time developing."

In fact, the two key points of the *Advice* were "stimulating demands" and "restraining production capacity."

The *Advice* regarded distributed PV stations as a key element in stimulating demand. The three measures to actively expand the market for PV power are, as provided in the *Advice*, "vigorously expanding the market of distributed PV stations," "gradually pushing forward the construction of PV stations" and "consolidating and expanding the international market." Compared to the international market, the Chinese domestic market boasts bigger opportunity for the PV industry. However, it has more room for distributed stations than for large-scale ground PV stations.

In August, 2013, the NEA published the list of the first "distributed station demonstration centers," covering 7 provinces, 5 municipalities and 18 demonstration programs. The *Notice on Developing the Demonstration Centers for Distributed Stations* issued by the NEA has pointed out that from 2013 to 2015, the total installed capacity of the 18 demonstration programs would reach 1.823GW, of which 749 MW was already under construction in 2013, and the rest would be completed before the end of 2015.

However, despite the good news mentioned above, I am more interested in measures that can ensure the healthy growth of this industry, such as restraining production capacity. All of these measures have been adopted to cope with the problems that must be solved if we are to develop the PV industry, and these problems are the root causes for the difficult situations encountered by most PV companies.

For example, an analyst from Shenyin Wanguo Securities Co. LTD pointed out that we can conclude from the *Advice* that, in fact, the government does not hope the PV industry continues to expand its capacity, but rather seeks to phase out backward capacity. The *Advice* provides that the conversion efficiency of monocrystal silicon in new PV manufacturing projects should be no less than 20%, that of polysilicon wafers no less than 18%, and that of thin-film solar cells no less than 12%. At the same time, it also provides that the overall electricity consumption of polysilicon wafers should be no more than 100 kWh per kilogram. For many Chinese PV companies, such standards are quite demanding because the conversion efficiency of even the best monocrystal silicon and polysilicon wafers manufactured in China cannot reach the goal of 20% and 18% yet. Conversion efficiency for monocrystal silicon cells remains between 19% and 19.5%. At the same time, companies producing polysilicon wafers consume electricity over 100 kWh/kg on average. Therefore, the aforementioned policies may indicate that the government is looking to phase out the backward capacity of all PV companies.

Indeed, the purpose of the *Advice* is to select the best and eliminate the worst in the industry. As it pointed out, "we will accelerate the M&A of enterprises, phase out the manufacturers with low quality and inefficient technologies, and develop the leading enterprises capable of conducting R&D and of competing in the market."

Accelerating the merger and acquisition process is another important measure for industrial restructuring. The *Advice* states that we should "encourage enterprises to conduct M&A through the force of the market." In this sense, as LDK Solar has already completed the first stage of M&A and is proceeding to the next stage, and Suntech Power is pushing forward its bankruptcy restructuring, the two companies can serve as models of the healthy development of China's PV industry.

Like Germany, the Chinese government also endeavors to develop the entire PV industry by nurturing technological innovation and raising overall competitiveness. The *Advice* pointed out that we need to "speed up the improvement of technology and equipment. By adopting the integrated programs of new energies, we can support R&D as well as the industrialization of highly efficient crystalline silicon solar cells, new thin-film solar cells, electronic-grade polycrystalline silicon, closed-loop circulating devices of silicon tetrachloride, high-end cutting machines, automatic screen printing machines, the plate type coating process, and key materials with high purity, etc. We will improve the technologies and equipment of PV inverters, tracking systems, power predictions, centralized monitoring and smart grids, etc., and improve the integration capacity of the PV generation system. We will support enterprises in developing new techniques in producing silicon and developing new PV products and technologies, and support backbone enterprises to build platforms for R&D and experiment in PV power generating technology. We will encourage higher education institutions and enterprises to cultivate their talents in the PV industry."

As far as I am concerned, the Chinese government has always regarded the PV industry as strategically important, and it has a clear goal to promote the healthy development of this industry. My only regret is that, although the government has clearly expressed its support for the R&D and industrialization of new thin-film solar cells, it has not formulated the supporting plans, policies and measures.

In fact, thin-film solar cells are the future of the Chinese PV industry, and will soon take the global lead. China has already acquired the technology, brands and capital in that field. For example, Hanergy already owns 7 thin-film solar cell technologies, three of which have reached advanced world standards, and four are top technologies in the field. In addition, the company has constructed 9 production bases for thin-film solar cells with a capacity of 3 GW, overtaking the biggest solar company in the US (2.5 GW) and becoming the world's largest solar company producing thin-film solar cells.

Of course, in time, the future of the Chinese PV industry will become clearer. With its current strategies, we can expect that the government will issue specific plans, policies and measures and turn them into concrete actions.

Direction for the PV Industry

Imagine a world where everyone owns a PV power station, and all the stations are connected together through a power line into a huge power grid. The electricity generated in your own station can meet your own demands, or can be sold to others through the grid. Likewise, you can also buy power, if you need to, from others, and you can even negotiate for a lower price.

Isn't this grid very similar to the World Wide Web today? Yes, and this will be the future of smart grids and distributed stations. These stations can be built on your roof, on the walls of your garden, and even on your clothes.

This will soon become our reality.

Integration Plan

Looking back, we may find that buildings and the construction industry have been the best allies of solar power. The use of heat from solar power proves those allies are necessary. Nowadays, on the roofs of many buildings, we can see rows of solar-powered water heaters; this has become a common scene in this energy-saving world. Developing the PV industry has become common in today's world. According to the *World Energy Report 2013*, Japanese BIPV account for more than 80% of all Japan's solar power generation. In Germany, this proportion is at 67%. However, in China, large-scale ground PV stations are the biggest contributor to the current installed capacity.

From the plans issued and measures adopted by the Chinese government, we can see the importance that implementing the new energy strategy promoting BIPV has taken.

The US and European countries have each carried out their own solar roof plans. In March 2009, in order to cope with the financial crisis, the Chinese government also put into force its version of the Solar Roof Plan, and carried out the Golden Sun Demonstration Project.

In the past, people would say that PV power does not only belong in western areas but also on rooftops. Now, we can extend this theory by saying that PV power does not only belong on rooftops, but also on the surfaces of buildings.

With the development of the second and third generations of PV technologies, thin-film solar cells can be used not only on rooftops, but also on surfaces of buildings. This has greatly improved BIPV. As a result, the area that collects sunlight can be increased many fold, and it is very easy to effect. In cities with high buildings, this technology undoubtedly has great significance.

As mentioned above, the area of China under construction has reached nearly 50 billion square meters, and its annual increase equals the sum of that of the US and Europe. By 2020, the area under construction will be about 90 billion square meters, and factory buildings will be specially fit for PV power generation. If we installed solar cells with a conversion efficiency of 10% on 10% of the rooftops and on 15% of the side walls, the total installed capacity in China would reach 1 billion kW.

Since 2012, the Chinese government has regarded the development of distributed stations as strategically important. In my opinion, as we focus on the development of distributed stations, we should also pay special attention to thin-film solar cell–based BIPV. Pushing forward their integration is a strategic measure for the energetic revolution.

Fine Birds Choose the Good Trees to Rest On

PV power is like a bird; buildings are trees. A BIPV is like a bird resting on a good tree. Their integration can bring about benefits in nine ways.

> Besides being used in solar stations in deserts and areas with rocky desertification, thin-film solar cells can also be used in BIPV. This is actually an important application of such cells.

First, BIPV is the best way to harvest solar power. Coal and oil are buried deep under the ground, while solar power comes from the sun. In order to generate power, we need to dig out the fossil fuel and harvest solar power from the sun. Buildings with PV cells are the best way to fulfill this task.

Second, BIPV is not as costly as building a new power station. If we popularize BIPV in China, then we can save a huge amount of money. Besides, BIPV can also increase the value of the original building. This will also bring about economic benefits.

Third, BIPV has improved the energy supply and consumption balance. With BIPV, a building can turn into a power station and establish a distributed power generation system in which people can use the power generated by themselves and sell the surplus. It can change the current power supply based on centralized power generation and distribution, making energy use more convenient and economic.

Hanergy is working with some renowned real estate developers to construct buildings with integrated PV stations. In this way, it can increase the added value of the buildings, provide residents with high investment returns, and reduce the use of fossil fuels all at once. Even e-commerce enterprises have started to march into this market. Jingdong has proposed small apartments called "Zero Carbon 2.0" with a floor space of 40 square meters. These apartments are priced at 1.2 million yuan. In cities where housing is already expensive, such as Beijing, Shanghai and Shenzhen, 30,000 yuan per square meter for such a smart room is a totally acceptable deal.

Fourth, BIPV has made the most of PV technology. Nowadays, with the rapid growth of PV technologies, especially that of second-generation thin-film solar cells, the integration of PV power and buildings has been popularized. Take Hanergy for example, which has produced soft, light and environmentally friendly thin-film cells which can be installed very conveniently on the walls of buildings.

Fifth, BIPV has the potential for a large domestic market. In 2020, if the installed capacity has reached 1 billion kW, then the market size of the PV industry would be 10 trillion yuan. At the same time, it will also drive the development of supporting industries and expand their market size 2 to 3 times. The overall economic gain would be 30 trillion yuan, which will contribute 4 trillion yuan in tax, thus providing jobs for millions of people.

Sixth, BIPV can boost both the energy and the construction industries. They are both pillars of China's real economy, and their integration can promote their restructuring and development. Right now, the construction industry accounts for one-third of the world's energy consumption. This industry's biggest challenges are energy conservation and emission reduction. BIPV makes it possible to improve both the energy and the construction industries. We can develop PV power on buildings, and develop new energy forms. Not only can we construct energy-saving buildings, we can also make these buildings energy-efficient by using solar power. As China is going through a period of rapid urbanization, in which lots of buildings are constructed every year, we can surely benefit from combining these two industries. It seems to me that this is a good opportunity for us to develop BIPV.

Seventh, BIPV can change the former investment structure. In the past, investments in the energy sector mainly came from the government and state-owned enterprises. Energy projects are usually big in scale and require a long time to come together, construct and start producing. Unlike fossil fuel projects, BIPV projects can be invested in by all kinds of enterprises, departments, and even individuals, because they are small and medium-sized projects that people need. They do not require too much investment or a long time for construction, and investors can gain benefit quickly without much risk. The distributed energy systems inevitably attract distributed investments from those who are both constructors and residents. As far as I am concerned, products which incorporate combined investments, consumption needs and even lifestyle trends will have great potential to flourish in the future.

Eighth, the distributed power systems will inevitably facilitate energy efficiency and energy conservation. The traditional system of centralized power generation and distribution is caught in a dilemma. On the one hand, if the tariff is high, then the cost of living will increase. On the other hand, if the tariff is set low, then power companies will benefit less and people will increase their power consumption excessively. However, under the system of distributed power generation, users construct the PV stations, then use the power generated by their own and sell the surplus to the grid. In my opinion, this system will serve as a control valve. Under this system, every user will try to improve energy efficiency and save energy so that they can benefit from selling the surplus.

Ninth, BIPV has paved a way for modernizing rural areas and agriculture. It will help make great breakthroughs in such things as upgrading the rural power grid and providing access to power for rural residents. In the vast rural villages, most buildings have very few stories but large rooftops, and many of them are built with a backyard. In addition, rural areas often use greenhouses to grow vegetables or flowers. Compared to urban construction, these rural buildings are more suitable for the installation of integrated PV stations. This has been proved by the rapid development of solar power in rural areas. The cost and investment for PV power may be higher than that of solar power heating, but as the technology of thin-film solar cells is upgraded and costs are reduced, PV power will one day prove to be the best choice for rural areas.

The aforementioned benefits are enough to prove the importance of BIPV. Even though we still have to build large-scale ground PV stations in the right places, we, with much consideration given to the big picture and the national situation, must focus on BIPV to implement the "strategy of new energy substitution."

PV: FUTURE PROSPECTS

Introduction

In China, there are legends and idioms about the sun and the people's attempts to utilize the sunlight. The Chinese have been dreaming of harnessing solar power for thousands of years. As we finally have entered the 21st century, it is time to make full use of this power. For the energy sector, the PV revolution embodies the ultimate replacement of traditional energies. In environmental protection, PV stands for the ultimate solution to pollution. For other related industries, like automobile, aircraft, and housing, PV represents the inexhaustible impetus.

PV matters a great deal to humanity. With it, humanity can say goodbye to the Oil Age and welcome the Solar Age. People will not need to fight each other for scarce coal and oil. Centralized industrial production as well as social organizations will become distributed, people will have a more open and inclusive mind, believing that resources are only used but not owned by them, and advocating sharing rather than monopoly.

Solar power has transformed people into the "PV generation," living under the blue sky and surrounded by green mountains and clear waters. They can achieve self-sufficiency in resources, and use clean and sustainable energy to provide inexhaustible power to automobiles, planes and ships. This is the "PV dream" of the Chinese people.

The PV Century

Why did the Americans dare to predict that by 2050, solar power will provide 69% of their electricity and account for 35% of their total energy consumption? Why, as the PV industry develops rapidly and coal and oil are phased out, aren't resources such as shale gases, nuclear power, biomass and tidal power the future of the energy sector? Why will solar power definitely replace fossil fuels in the future? It is because the 21st century will be the solar century—the PV century. This is the impact of the PV industry on the energy sector.

Technology is Advancing Rapidly

Based on our experience with technological transformation and the development of PV, we may conclude the following.

First, technology transformation will continue to make the PV industry progress. PV technologies have developed for more than 40 years. As early as 1969, the world's first solar station was built in France and the PV industry was born. Since then, technological breakthroughs have been made. It is worth noting that Germany also made technological progress after France. In April 2011, Bosch Solar Energy achieved a 19.6% efficiency for large-area PERC (Passivated Emitter and Rear Cell) solar cells, which were the most efficient solar cells in the world at that point. Besides European countries, the United States has also made breakthroughs in this field. In July 2011, First Solar successfully produced cadmium telluride cells with an efficiency of 17.3%.

As Europe and the US have made huge efforts in experimenting with solar energy, technologies are advancing faster than anyone could have expected. When we founded the China New Energy Chamber of Commerce in 2006, the cost of solar power generation was 3–4 yuan/kWh. In 2009, less than 3 years later, we had reduced the cost to less than 1 yuan/kWh, and the current cost is about 0.5 yuan/kWh.

PV technologies today are advancing rapidly and becoming more and more mature. We will make bigger breakthroughs in the technology of monocrystal silicon, polycrystalline silicon and thin film, and we will also improve the conversion efficiency of solar cells as well as equipment manufacturing and system integration technologies. We are confident in predicting that in the near future, we are to witness tremendous transformations and a developed PV industry with lower power generation costs, as technologies continue to improve and commercial applications grow in popularity.

Second, new technologies will foster big companies, who will in turn drive the development of new technologies. As thin-film technology has advanced, many companies that rely on it have risen in the US, South Korea, Japan and in Europe.

Other examples include computer technology giving birth to first-class computer companies, e.g., IBM. IBM, with its investment and R&D in personal computers made the computer an everyday item for ordinary people.

However, companies can be doomed if they go against new technologies. Motorola is an example in the communications industry. Before the 1990s, Motorola had been the pioneer in the communications sector. In July 1969, the Motorola-made wireless device sent back the first greetings from astronauts on the moon. However, in the 1990s, when other communications companies were transforming from analog communication to digital communication, Motorola failed to catch up with the trend and consequently was bought and marginalized by Google.

The End of Fossil Fuels

The new technology revolution and new Industrial Revolution mark the rise of solar power and the advent of the PV century; on the other hand, they also indicate that fossil energies will soon be phased out.

Fossil energies will be phased out because, on the one hand, industrialization's consumption of fossil energies is dramatically high. Emissions from burning fossil fuels also cause global environmental pollution. On the other hand, the development of fossil fuel substitutes, especially renewable energies, has made the replacement of fossil fuels possible, and the development of PV has made this trend irreversible.

The exhaustion of fossil fuels is eminent, which can be demonstrated by the two major oil crises after the Second World War. The two crises, which lasted from 1973 to 1974 and from 1979 to 1980, resulted in oil price hikes on the international market (for example, the price soared from USD 13/barrel in 1979 to USD 34/barrel in 1981), and severely impacted Western countries.

Meanwhile, the mass use of fossil energies has given rise to pronounced environmental problems. Unusual phenomena in the world such as global warming, the greenhouse effect and sea level rise are all caused by the use of fossil fuels.

Right now, as fossil energies are on the verge of exhaustion and are placing the environment in great peril, the development of renewable and clean energies enables mankind to restructure and optimize the energy mix. According to the *BP Statistical Review of World Energy 2013*, the power generated by renewables in 2012 increased by 15.2%. When the growth of the world's total power generation slowed down, renewables other than hydropower contributed to 4.7% of the total generation mix. In the same year, the growth rate of CO_2 emission from energy use was slower than that of 2011. Reducing its usage of coal will help America reduce CO_2 emissions to the level of 1994.

The development of PV has accelerated the replacement of fossil fuels by new energies, which is manifested in two respects: first, some countries have proposed replacing fossil fuels with PV; second, consumers are willing to pay more for PV.

Many countries have specific plans to replace fossil fuels with PV, and the United States is a typical example. In 2008, Ken Zweibel and two other American scholars devised a grand scheme: by 2050, solar power will provide 69% of electricity and 35% of the total energy consumption (including transportation-related energy consumption) of the United States. They predicted that energy prices for the same period will be 5 cents/kWh, which is approximately the same as the price now. If wind energy, biomass and other energies are fully developed, then 90% of the energy supply and consumption in America will have been derived from clean energies by 2100. Currently, the shale gas frenzy may have slowed down the development of other new energies, but the U.S. is still a major player in solar power development and utilization.

In terms of the consumer, in 2009 the International Environmental Protection Organization and Association announced that, according to an Ipsos survey conducted in ten cities in China, about 80% of the interviewees agreed that the use of coal would heavily pollute the air, and that they are willing to pay a 19% price hike to buy clean energies like solar power and wind energy.

A prerequisite to the above predictions is the current development of the PV industry. If development evolves into revolution, then the replacement of fossil fuels by PV will not take 40 years, but only 35 or 30 years, or even less.

> The PV industry in China has international competitiveness in matters of capital, talent, technology and market, but it also has four shortcomings. First of all, the PV industry has not developed in an orderly manner: the production capacity of polycrystalline silicon increased rapidly with unbalanced supply and demand. Second, the PV industry is heavily dependent on foreign markets, and the domestic market needs stronger policy support. Third, the future of the PV industry is challenged by the lack of innovation and technology development. Fourth of all, protectionism is on the rise in the global market, leading to the deterioration of our external environment for the sake of industrial development.

The Irreversible "Ultimate Substitute"

Statistics suggest that, even though fossil energies like coal and oil are still playing a predominant part in the energy mix, they will eventually be replaced by renewables, and this trend is irreversible.

New energies such as nuclear power, wind power, biomass and tidal power are reshaping and optimizing our energy mix. According to the *BP Statistical Review of World Energy 2013*, oil is still a globally dominant fuel, accounting for one-third of world energy consumption. However, according to data reaching as far back as the first BP energy review in 1965, the share of oil in the world energy consumption mix is at a record low. The use of coal is growing faster, but compared with historical data, coal is among the fossil energies with the fastest declining growth rates.

The substitution for fossil energies by renewables is imminent. In only 10 years (2002–2012), the percentage of renewables—except hydropower—in world energy consumption grew from 0.8% to 2.4%. The substitution may not be huge in total size, but nonetheless significant in growth rate.

However, the prospects for other renewables such as nuclear power and biomass are not as bright as that of solar power. According to the *BP Statistical Review of World Energy 2013*, power generation from nuclear power decreased by 6.9%, its biggest recorded decline for two consecutive years. As the output of biofuel in the United States decreased by 4.3%, the world's biofuel output also dropped for the first time since 2000. In addition, due to unstable power generation and uneven resource distribution, the development of wind power has met setbacks in recent years.

Nuclear energy is subject to limited uranium resources; biofuel requires a large number of crops, which is not cost-effective; tidal power also requires a lot of resources and it entails a huge impact on the environment. As for shale gases which have been advocated by the Americans, it belongs to the category of fossil fuels, not to that of clean and renewable energy.

Considering all the relevant factors, only a new type of energy meeting all the following criteria may serve as the ultimate solution to the energy problems of mankind. First, the substitute must be clean and renewable; therefore shale gases are not an option. Second, the substitute has to be distributed evenly, so tidal and wind power are out of the discussion. Third, the substitute's production and utilization process must be safe and controllable; therefore nuclear power is not an option, either. Fourth, the substitute has to be cheap and its cost controllable, so coal and oil are excluded.

PV energy meets all the above criteria. First of all, solar power is derived from the sunlight, posing no risks in matters of geological development and irreplaceable development. PV is thus a genuine renewable energy. Second of all, solar power is distributed in an even and balanced manner, leaving no idle equipment and giving every region equal access and chance to develop solar power. Third of all, PV poses no threat of a leak, and the production of PV is pollution-free. Last but not least, despite a short development history and imperfect industrial base, the PV industry has high power-generating efficiency.

Therefore, replacing fossil fuels with solar power is not one of the choices, but the only choice. From a practical perspective, solar power is the inevitable trend in the long run. When is mankind going to completely phase out fossil energies and apply PV technology to the globe as a whole—it is only a matter of time, technology, location and politics.

> Nowadays, technological progress keeps lowering the cost of solar power generation while the cost of power generation from fossil energies is on the rise. When the cost of solar power becomes equal to that of fossil energies, solar power will replace fossil fuels on a large scale.

Replacing fossil fuels with PV is absolutely feasible. As mentioned before, the cost of PV power generation has dropped to 0.5 yuan/kWh. This figure is significant, because if we consider the cost of environmental damage caused by fossil energies, the total cost of fossil energies would be higher than 0.5 yuan/kWh. It is time for us to massively replace fossil fuels with PV!

However, it is worth noting that this substitution is not an interim one, but the ultimate one. This means that when we complete the replacement, there will be no other substitute for solar power. Men will bid farewell to coal and oil, oil wars and nuclear leaks, and enter an eternal solar age.

New Landscape on Wheels

When talking about solar energy, people immediately think of solar water heaters on rooftops. Is that all that solar energy has to offer?

Why did Li Shufu, Chairman of Zhejiang Geely Holding Group Co., Ltd, and I decide to install a solar sail on a car roof to generate electricity? Why not fuel future planes with solar energy?

The influence of PV on related industries includes but is not limited to the issues mentioned above. PV has a bearing on ground power stations, buildings, transportation, and even common consumer goods like lamps and tents.

The Burgeoning PV Stations

Back in 1893, Chicago was preparing for a World Expo. Founded in the 1820s, this international commerce and trade event grew along with industrial development, incorporating elements of science, art, and life into exhibitions. At that time, two great minds in science—Edison and Tesla—both set their eyes on the World Expo and prepared for a showdown on this international stage that was the focus of the world.

The reason for their competition was that the organizing committee was looking for lighting equipment to illuminate the entire venue, and the candidates were Edison's DC solution and Tesla's AC solution. As a result, in January 1893, more than 90,000 AC-lights illuminated the venue at the opening of the expo.

In 1895, equipped with Tesla's AC technology, the world's first hydroelectric power plant was built in Niagara Falls. Electricity generated was transmitted to Buffalo, 35 kilometers away from the plant. This event implies that AC vanquished DC, making the latter an outdated technology.

DC and AC wrote the prelude to the second Industrial Revolution, and the application of AC gave birth to ground stations, which will be further impacted by PV, the forerunner of the third Industrial Revolution.

The impact is mainly of two kinds: on the one hand, stations powered by clean technologies, e.g., PV, will gradually replace traditional fossil-fuel-powered ones; on the other, since solar energy is extensively distributed, future power stations will also be run in a distributed manner.

In terms of the current structure, traditional power plants are still the major electricity suppliers. According to Bloomberg New Energy Finance (BNEF), despite the fact that statistics for different regions vary in terms of power generation, fossil fuels remain the dominant source of power generation. Thanks to the incentive policies issued in Europe, the US and recently in China, electricity generated by renewables—except nuclear power—increased by 64% from 2006 to 2012. Wind and solar power grew the fastest. Indeed, with huge cost cuts, solar power generation has been developing rapidly.

Meanwhile, the investment market for new forms of power plants—and especially solar plants—is also booming. BNEF statistics suggest that PV and other renewables enabled global investment in clean energy to score a 17% compound annual growth rate (CAGR) from 2006 to 2012, highlighting the future of the industry.

From the perspective of cost, PV plants are already more competitive in price than traditional power plants. The advantage is not so prominent if we simply compare the costs in their absolute number, though. However, if we take the plummeting cost of PV systems and include reduction in carbon emissions, the price of solar power generation can compete with that of coal-fired power generation. As PV technology continues to develop, the price edge that solar plants have will become more pronounced.

The trend shows that 2020 may be the turning point, where new energy power stations replace traditional ones. By then, solar power generation will be the second largest growth pole apart from wind power generation, and 427GW of solar power capacity will have been newly installed. It is worth noting that such analysis is made in line with, rather than in advance of, the steady progress of PV revolution.

In the meantime, as mentioned above, solar power is distributed extensively and unevenly, so future power plants are likely to be operated in a distributed manner rather than a centralized one. Distributed power generation often refers to highly-efficient and reliable power-generating units, which are small modules with a generating power ranging from several kilowatts to hundreds of megawatts (some people set the range between 30 and 50 MW) distributed near users.

The beauty of distributed PV power generation is its low cost and high efficiency. Building a traditional power plant requires a one-time investment of several billions and a construction period of 5 to 15 years, entailing huge fund risks.

Distributed power plants require less investment and can be quickly put into operation. If we replace traditional power plants with PV stations, we can provide customers with quality and reliable electricity without worrying about the complicated structures of traditional plants. At the same time, distributed PV stations can break up monopoly and factor in competition so as to increase the utilization of the energy. This power generation model can also be applied to wind power or biomass.

BIPV

After discussing PV's influence on power plants, let's look at the impact of the PV revolution on the construction industry.

Figure 6-1 The sectors of PV application in the world between 2008 and 2012

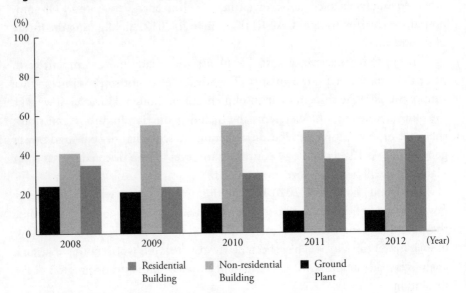

Source: Global New Energy Development Report (2013)

When speaking of the impact of solar power on buildings, ordinary people often refer to solar water heaters on rooftops. This is actually a misunderstanding. Those heaters are only an application of the photo-thermal effect of solar power, not an application of PV technology. The impact of PV on buildings is in BIPV.

BIPV is the integration of architecture, technologies and aesthetics, and does not deface buildings as solar water heaters do.

In June 2012, IKEA declared its strategic partnership with Hanergy: all IKEA stores in China will be equipped with solar PV panels to promote the energy conservation and emission reduction project of the IKEA Group. More than half the electricity used in IKEA buildings worldwide is produced by clean energies like wind and solar power. 250,000 solar panels have been installed on IKEA buildings around the globe. IKEA owns and maintains 80 wind turbines. Taking IKEA Beijing as an example, the installed power capacity of its rooftop solar panels has registered 416.24kW, reducing annual carbon emissions by 426 tons.

Wu Changhua, the Greater China Director of The Climate Group, an international non-profit organization, has said: "China is going through a unique development period and it has to deal with unbalanced and urgent challenges and opportunities, among which attaining clean energies and emission reduction quotas are the top priorities requiring immediate action of leading enterprises. We appreciate the timely alliance between IKEA and Hanergy, and meanwhile, we expect that more companies and all walks of life will actively get involved so that we can join hands to realize the beautiful Chinese Dream featuring a clean, green and low-carbon life."

As for the BIPV industry, China, in the current context, can carry out pilot projects before mass promotions. For industrial, business and residential buildings both in place or under construction, the government can slash or exempt deed tax, property tax or other taxes as long as the buildings are BIPV-renovated. The government can also take BIPV installation as a prerequisite for the approval of new buildings, and thus encourage self-consumption, feeding in the grid and paying for the net volume of electricity transmission between the grid and BIPV systems. In addition, such policies as loan priorities and preferential interest rates will motivate BIPV enterprises.

PV-Based Transportation

The PV revolution will bring disruptive change to the automobile and transportation industry.

Cars are regarded as the "machines that changed the world": machines whose emergence and development created the auto industry, invigorated the oil industry and changed people's way of work and life. The whole of society now sits on wheels. In 2011, global automobile production was 80.1 million units, with a year-on-year growth of 3%, and global car ownership was 1 billion. During this period, the automobile industry also became an economic pillar for America, Japan, Germany and other developed countries. There are cars everywhere, making the Earth a world of cars.

However, the increasing number of cars comes with two consequences. One is that rising automobile production leads to greater oil consumption. Global oil consumption reached 88 billion barrels a day in 2011. Another is that increasing oil consumption brings about more serious pollution from auto exhaust. Thirty-five percent of the world's carbon emissions is derived from oil consumption, of which cars are the major source. Though we recognize the convenience brought along by cars, we should also be aware that reducing oil consumption and addressing car-related pollution have already become major challenges calling for immediate action.

Replacing oil with electricity to fuel cars has been a trend since long ago. Actually, electromobiles were invented earlier than motor vehicles. In 1886, the German, Karl Benz, "father of cars," invented the first vehicle powered by an internal combustion engine. However, back in 1834, the American Thomas Davenport had already invented the first DC-motored electromobile, the first of its kind in a real sense. Then electric vehicles (EV) soon substituted for steam-powered ones, thanks to their speed and driver-friendly operation. However, they failed to gain popularity because of battery restraint. In 1908, cars powered by internal combustion engines were invented by Henry Ford for mass production and soon took over the market. Their popularity boosted the development of the oil industry and lowered oil prices, which in turn reinforced the popularity of those vehicles. EVs were soon abandoned.

Yet six decades later, EVs seized the opportunity of the two "Oil Crises" of 1973 and 1979 to challenge traditional cars twice. However, they failed again because of chronic flaws: low cost-effectiveness, inconvenient recharging, and short ranges.

Today, the rise of the PV revolution has granted EVs new opportunities. And here I want to share a story.

Recently, I traveled abroad with national leaders as a representative of Chinese entrepreneurs. On the flight back, I happened to sit next to Li Shufu, president of China's first private automobile company, Jili. We talked and both learned something new from each other. Of course, our conversation was centered on the topic which concerned Li the most: vehicles. He told me that every kilowatt-hour of electricity an EV consumes allows the car to travel 8 to 10 kilometers, and I was amazed by the fact that so little electricity was enough to power a car. Relating to Hanergy's technical advantages, I replied: "Covering cars with thin-film PV can create a solar power system with a capacity of 2–3kW, which ensures a range of more than 100 kilometers." Li was impressed as well because he had not known that thin-film PV was so developed.

The shocking information we learned from each other inspired innovation. We immediately agreed to cooperate in designing solar-powered vehicles.

Meanwhile, we also came up with an interesting idea—to install an adjustable solar sail that can increase lighting as needed, generate more electricity and allow the car to travel further. We have already begun the upfront R&D of such cars.

In developing and promoting the use of EVs, insiders make calculations based on the economic indicators of EVs. If an oil-fueled car consumes 10 liters of gas every 100 kilometers, and suppose every liter costs 8 yuan, traveling 100 kilometers would cost 80 yuan. If road conditions and other variables are kept the same, an EV will consume 12 kWh of electricity every 100 kilometers, supposing the power tariff is 0.5 yuan/kWh, the cost for an EV to travel 100 kilometers is only 6 yuan—not even one-tenth that of an oil-fueled car. However, the calculations are only applicable for plug-in EVs. With current PV technologies, it is possible for us to depend on thin-film PV covering the vehicle to generate electricity. If such PV cars can be widely put into use, it will mark the third challenge initiated by EVs to their oil-fueled rivals.

Concerning the battle between PV-powered and oil-fueled vehicles, let's just wait and see who will be the ultimate winner and dominate the market. The final result is obvious and the traditional auto industry will be subverted, triggering a huge industrial transformation involving billions of yuan.

With the application of the PV revolution in the auto industry, similar energy substitutions in trains, planes and ships should also be within our reach. It sounds like a fairy tale, but in a world making such rapid scientific and technological progress, we have already seen the beginning this fairy tale in reality.

On July 8, 2010, "Solar Impulse," the solar aircraft flown by the Swiss explorer André Borschberg, successfully completed the first test flight of its kind, setting a record for non-stop flight of 26 hours and 9 minutes. Another Swiss explorer, Raphaël Domjan, cooperated with Swiss and German manufacturers to create a solar catamaran in 2009. The catamaran started its journey in Monaco in September 2009, traveled a total of 60,000 kilometers in a year and a half and finished in Monaco on May 5, 2011. On June 6th of the same year, the world's first solar train was put into operation.

Household PV Appliances

The influence of the PV revolution is not limited to the power industry, the building industry or the transportation industry. As such a revolution intensifies, the impact will also extend to goods for everyday consumption which are closely connected with people's economic life. Household appliances are the number one example.

The extensive application of electricity in human life is an important characteristic of the second Industrial Revolution, and its application in consumer goods catalyzed the rapid development of household appliances in the industrial structure.

Currently, widely used household appliances include fridges, washing machines, air-conditioners, TV sets, telephones, computers, lights, electric cookers, etc. All these constitute material essentials in people's lives; indeed, we can hardly imagine a life without them.

The household appliance industry is well-developed. To take the Chinese market as an example, China Market Monitor Co., LTD, a third-party research institution, predicted that the household appliance market in China would be worth 1,188 billion yuan in 2013, with a year-on-year increase of 3.2%. And appliances in the broad sense, such as computers or telephones, were not included. Otherwise, the size of the domestic market would be even bigger.

The further development of the PV revolution will exert a profound influence on the household appliance industry, an influence backed-up with evidence.

Actually, as we are embracing the third Industrial Revolution, the household appliance sector has already been reshaped by Internet technologies, which are also part of the revolution. At present, profound changes are taking place in the Internet-friendly household appliance industry, e.g., "the cloud appliance," with successful efforts in fridges, washing machines, air-conditioners, etc. At the same time, with the help of Internet technologies, domestic tycoons like Media and Hair have achieved the mobile interconnection of home appliances.

The PV revolution will improve household appliances in a holistic manner. On the one hand, appliances that used to be powered only when plugged in will now have access to electricity generated from solar power: solar cell phones, solar computers, solar printers, solar lamps, solar induction cookers, ovens etc. On the other hand, the PV revolution will broaden the range of appliances, even taking non-appliances into the sector: solar tents, solar watches, solar tables and solar tea sets, etc.

The booming household appliance industry over the years has made energy consumption and pollution problems even more acute. To address relevant issues concerning industry development, the Chinese government has been focusing on setting standards so as to increase competition in promoting energy conservation and pollution reduction. Appliance industry analysts at S&P Consulting believe that higher standards for energy efficiency would phase out many products, cut the number of producers and reshuffle the industry.

The PV revolution will address issues of energy consumption and pollution for good by strongly impacting the household appliance industry. It will also introduce more investment into the industry as 3D and smart household appliances become a reality.

> The solar industry belongs to the real economy, and matters a lot in restructuring the economy, because the development of the solar industry can boost multiple other industries, including high-end equipment manufacturing. It will definitely become a pillar industry, emerging as another growth pole besides urbanization that can rapidly and effectively propel the domestic economy.

A Green Home Is Not a Dream

China missed the first Industrial Revolution, and only got involved in the last 30 years of the century long second revolution. We were unfortunate but at the same time we were also lucky in all our misfortunes.

It is easy to interpret our experiences as misfortunes. But by "lucky" we mean: fossil fuels brought about two Industrial Revolutions together with a severe eco-crisis. Since we were not fully involved in the first two revolutions, yet we are ready just in time to catch up with the third, we can grasp the opportunity of developing PV and realize our dream of building a "green home."

Ecological Crisis Looms Just Ahead

"If civilization develops spontaneously rather than consciously, it will only leave humans with deserts." One hundred and thirty years ago, Marx warned people of the Industrial Revolution that was developing by leaps and bounds. Sadly, although this quote was highlighted in *Marx and Engels: Collected Works, Volume 1*, those who led the revolution did not heed its profound meaning. Therefore, human beings keep on walking further and further along the road of "desertification," seriously damaging the eco-balance.

Through his production and lifestyle, man is accountable for Earth's ecological problems, especially those caused by the excessive and inappropriate use of fossil fuels.

Taking the greenhouse effect as an example, the scientific community believes that fossil fuel consumption leads to GHG emissions. The increased GHG concentration then enhances the greenhouse effect and causes global warming. Since 1860, the world average temperature has risen by about 0.4 to 0.8 °C. According to an IPCC report on future climate change, CO_2 is the major component of GHG, and 90% of manmade CO_2 emissions are caused by fossil fuel consumption.

China's experience with development is inspiring as well. What developed capitalist countries achieved in 200 to 300 years of industrialization only took us 30 years of reform and opening up. However, for that, there is a price to pay. Due to lack of natural resources, we are becoming increasingly reliant on imports from international markets.

Under such circumstances, China has prominent environmental concerns. The total amount of GHG emissions in China has been enormous and the growth is so fast, imposing greater challenges to addressing climate change. China's cumulative CO_2 emissions between 1850 and 1990 only accounted for 5% of the world's total and little pressure on the environment then. Now, we are already ranked the world's top CO_2 emitter, contributing 24% of the emissions worldwide. As the IEA predicts, China will overtake the EU in cumulative emissions to become the second largest emitting country in the world in 2035.

Environmental issues trigger discussions on many major topics, such as economic development, energy utilization, international competition, people's well-being and the destiny of human society, etc.

For developed capitalist countries, outsourcing heavily polluting industries like manufacturing to developing countries has already become force of habit. For example, in the last 40 years of the 20th century, after achieving skyrocketing economic development, Japan transferred more than 60% of its heavy industry overseas due to labor shortage, as well as in an attempt to control cost and improve the domestic environment. During this gradient transfer of industries, America has also transferred 40% of its highly-polluting industries outside the country.

Fifty years ago, developed countries were able to shift the environmental crisis to the rest of the world. But today, it is impossible for China to follow suit as Internet technologies are highly developed and everyone attaches great importance to environmental protection.

Therefore, we are faced with the arduous task of tackling these environmental issues. In 2012, the 18th CPC National Congress made comprehensive arrangements for ecological development and incorporated the promotion of ecological progress into the CPC Charter. It was a first in the development or ruling history of any political party. The academic community regarded this initiative as a leap in cognition, an innovation in theory, and a breakthrough in practice. It is a milestone in ecological development which set China on a new march toward sustainable development. It is also a sensible act considering the grim pollution levels in China.

Solving environmental issues in China may be the dream and cause of a whole generation of people. According to estimated statistics from the relevant industries, total expenditures for water treatment in China will register 2 trillion yuan; when it comes to air pollution, the data are similar. So, altogether, treatment costs 4 trillion yuan, but the national fiscal revenue in 2012 was only of 11.72 trillion yuan. What's worse, although we can calculate the cost for tangible items, the intangible value of time spent is hard to predict, and it may take a whole generation's time before our final goal is attained.

The PV Revolution is the Way Out

As a bystander at the first Industrial Revolution and a latecomer to the second one, China may have been spared from the development model of "pollution first and treatment later." Thirty years of industrialization did leave behind some ecological problems, but they are not serious enough to get us mired. If we can seize our chance with the third Industrial Revolution and quickly move into the new energy age, we will probably pay less than Western countries in economic development, energy utilization and environmental protection, thus countervailing the negative effects of the past two Industrial Revolutions.

Fortunately, we are living in a society where technologies evolve with each passing day. Thanks to PV development, China may spend less time than America getting rid of pollution and embracing a clean environment.

The PV revolution favors environmental consciousness. First, solar power is clean and renewable, generating neither poisonous gases nor hazardous wastes. Second, solar power generation is distributed, so it causes no water pollution or noise pollution unlike major construction projects. Third, the urbanization of China is in full swing. The integration of urbanization and the PV revolution will prevent us from making the mistake "pollution first and treatment next."

Of course, the energy substitution brought along with the PV revolution is only a physical measure to prevent ecological damage. If we want to eradicate environmental problems for good, we have to become environmentally conscious, which is the fundamental solution. In this regard, the "zero carbon home" and "zero carbon community" promoted by foreign countries are good examples of an environment-oriented awareness. The "zero carbon suburb" of Sutton in South London is a good example.

Designed by the famous British eco-architect, Bill Dunster, the community was built on waste-backfilled ground, with every detail focusing on the recycling of sunshine, waste, water, air and lumber. The Beddington eco-village in London is a community taking up one hectare, and is home to hundreds of buildings, thousands of square meters of office area, an exhibition center, a kindergarten, a community club and a football field. Dining in a zero-carbon home, you will find that leftovers are turned into raw materials for heating and power generation. The sloping rooftops of the "zero carbon community" look like waves with solar panels installed on southern rooftops, which serve as the main energy source for the zero-carbon buildings. The facade skin of the buildings comprises several layers (from outer to inner): cement fiberboard, natural insulation materials, bracing structures for exterior walls, and high-density gypsum plaster for interior walls. Such design incorporates all kinds of natural energies, be it the sunshine or rainfall, and transforms them into electricity and drinking water as well as water for other usages.

The Zero Carbon Community, with solar power as its energy source, has a whole package of designs and solutions to minimize the negative impact of human activities on the environment. And in practice, we can break down the idea of a "zero carbon community" into individual solutions targeting water, solar power, and household refuse. Without mature conditions, these individual solutions can also protect the environment. Replacing our most polluting energy with PV dramatically alleviates environmental pressure. It is the substitution of traditional energies with PV and the "zero carbon community" that enable China to dream of an eco-civil development.

PV Contributes to the Chinese Dream

Why do we say that the PV revolution goes hand in hand with the Chinese Dream? The answer to this question explains the significance of PV in China—the PV industry in China emerges right as Chinese PV companies are starting to dream and drive. "The Chinese PV dream" is not only an important component of the Chinese Dream; it is most importantly a cornerstone of the Chinese Dream.

PV Supports the Chinese Dream

The PV revolution is an important component of the Chinese Dream. It provides indispensable physical support to the dream.

After 30 years of reform and opening up, China now stands at an economic crossroads. Our development mode is shifting from relying on investment and export to relying on domestic demand, and our growth model is turning from an extensive one to an intensive one. The ultimate goal of such transformation is the sustainable development of our economy.

In recent years, we indeed have uncovered many shortcomings in our extensive economic growth, including the imbalance between the quality and scale of economic growth, development at the cost of environment, lack in core technical competitiveness, and heavily indebted local governments. If these things remain unchanged, our economic development will be undermined, which will further jeopardize the realization of the Chinese Dream.

To prevent further deterioration, Premier Li Keqiang of the State Council required that measures be taken to regulate economic activities, restructure the economy, streamline government, plan out urbanization and promote economic transformation.

To be more specific, by regulating economic activities, we mean regulating the bad practices which give false signs of economic development. These practices include the oversupply of resources and low efficiency at the micro level. Efficiency here refers to the ratio between input and output. Currently, both the output and input of China's economy are high, which is quite wasteful. Loss derived from low efficiency is seen as an internal waste in economic growth. There are many ways to achieve economic growth and technological advancement is the most important resort.

The PV-motivated energy revolution harmonizes perfectly with Premier Li's demands for the regulation of economic activities.

PV stations are not subject to the regional environment. The construction investment is not as huge as that for traditional power plants. Since PV stations are more cost effective, many inefficiencies seen in traditional plants do not occur. The PV revolution can also boost industrial development in a way which represents the future development trend. Taking the automobile industry as an example, it is commonly admitted that new energy cars will eventually replace vehicles fueled by fossil energies. Amid the many solutions of new-energy automobiles, PV vehicles, with their advanced concept and energy efficiency, are destined to be the future. The PV revolution, other energies and the industrial revolutions influenced by PV will definitely play an important part in regulating the economic activities of China in the near future.

By *restructuring the economy,* we mean balancing the percentage of three particular industries (structural optimization of agriculture, pillar industries, and traditional/modern service industries) as well as to improving the fiscal and taxation system.

The PV revolution will play an active role in economic restructuring. The technology-led PV revolution will breed many energy companies based on PV and other supporting technologies, and the birth of such companies will optimize the energy structure of China. At the same time, if the PV revolution can be widely promoted, industrial companies with high energy consumption will be somewhat relieved from pressure. PV-inspired industrial transformation will lead to the upgrading of other industries. The "zero carbon community" mentioned above shows how new concepts and tremendous changes are introduced to the building industry.

Urbanization refers to the process by which rural people become urban people. An important indicator of urbanization is the urbanization rate: the ratio of permanent residents in a region against total local population. The world urbanization rate is already over 50%. The *Outline of the National 11th Five-Year Plan* identified the construction of urban clusters as a major means of promoting urbanization. The 12th Five-Year Plan again emphasized that based on big cities, we should cultivate influential city clusters and promote the coordinated development of both cities and towns, with a focus on mid- and- small-sized cities.

Regarding urbanization, there is an important consensus on the necessity of integrating urbanization and ecological development. The current technologies suggest that ecological development is not merely a concept. More importantly, benefiting from PV-inspired technological advancement, ecological development is a crucial way to treat pollution caused by energy consumption. Therefore, the PV revolution and urbanization are two interdependent concepts.

Transforming the development mode refers to the economic transformation and upgrading of China. China's economy has boomed over the years, but meanwhile, it faces pressure from both home and abroad. In terms of the external environment, the world is speeding up its economic restructuring and a new pattern is taking shape. Developed countries attach great importance to real economy, and expedite developing new energy and new materials to take control of the future of science, technology and industries. As for our internal environment, profound conflicts and problems in industry are more and more prominent. We are facing imbalance in our industrial structure, overcapacity in certain industries, overdependence on investment and export, lack of innovation, core technology and domestic brands, and bottlenecks in energy development. In a word, China desperately needs to kick off economic restructuring and industrial transformation.

As for China's economic upgrading, backbone enterprises have to innovate and set higher targets in order to upgrade themselves, which is even truer for PV companies with the ambition to lead in the international arena. The international financial crisis has accelerated progress in science and innovation, propelling industrial revolution and restructuring in the world. Developed countries are accelerating strategic adjustments for science, technology and industrial development, trying to make breakthroughs in green and low-carbon technologies. In 2012, America launched a green economic recovery program and the EU started green technology R&D. Both measures were intended to enhance competitiveness and take control. As for China, although it has world-leading thin-film PV technology, on the whole, its crystalline-silicon producing capacity remains huge. Our PV industry is now at a crucial stage that can only be sustained through transformation. Only then can PV technology contribute to the Chinese Dream.

Apart from its economic contribution to the Chinese Dream, the PV revolution has in itself great spiritual value. Values like "sharing" and "harmony" are advocated in this revolution and they will replace values like "possessing" and "controlling" which took root during the first two Industrial Revolutions. Rooted in the third Industrial Revolution, which is triggered by the PV revolution, the great Chinese civilization, featuring modesty and fairness, will thus further promote other civilizations in the world along with China's economic and social development.

> The PV industry has to upgrade its industrial structure if it aspires to rise. First, we must identify the direction of the Chinese PV industry's technological development, because the direction determines whether the PV industry can achieve industrial restructuring and sustainable development. Second, we must give more support to BIPV as well as other PV applications.

PV Illuminates New Rural Areas

July 11, 2013 was a special day for the village of Fenghuangzhai in the town of Xinjie (Suizhou, Hubei). The first batch of power generation pilots for rural areas was installed in Fenghuangzhai and the village achieved on-grid power generation. Every year from now on, 20 households in the village will generate 60,000 kWh of electricity, allowing for economies equal to 20 tons of standard coal, and generating economic benefits of 52,200 yuan (based on an electricity tariff of 0.87 yuan/kWh).

PV technology is not only applied in rural areas of central regions, but also in more remote areas. For instance, the countryside in Tibet lacks energy. Large- and mid-sized power plants are concentrated in populous cities. By contrast, only a few small power stations are built in counties and villages. For a long time, people have been using animal manure, lumber, and turf as fuel, putting greater pressure on the already fragile environment. Power shortage is a major roadblock to the development of rural areas.

Some rural areas in Tibet have resolved the above issues with PV power generation. By harvesting solar power, Tibet has made breakthroughs in terms of communications, broadcast TV power supply, and access to non-electrified regions, and it has successfully launched special solar programs, becoming the first and largest administrative region at the provincial level promoting special solar programs.

Rural areas in the Ningxia Autonomous Region are also pioneers of PV solar energy utilization, and the town of Yongkang of the Shapotou District in

Zhongwei City is a typical case. In June 2011, 1,000 households in Yongkang Town installed solar panels with a power of 3.2kW on their rooftops, integrating into the urban grid of Zhongwei City via the 511 west optical line of the substation (35KV) in Xitai Town. The project implemented on-grid operation, equipment configuration, grid transformation, tariff estimation, and BIPV construction. After completing the project, local farmers can now use solar power for lighting, cooking, water-heating and showering.

The PV revolution has improved the living standards of local people, but also changed the local production model.

On March 1 2013, a PV-integrated 8.4 hectare vegetable greenhouse in the village of Zhangjiayingqian of the town of Daotian (Shouguang, Shangdong) went through all approval procedures for grid integration, and was successfully integrated into the State Grid. This demonstration park can generate 1.5 million kWh of electricity annually. Besides toughened glass and PV panels, the greenhouse's rooftops are equipped with waterproof and photic insulation sheet. PV panels on the rooftops protect vegetables from ultraviolet rays and pests, to improve the quality and amount of produce. While generating electricity, PV panels ensure and facilitate the photosynthesis of plants.

The PV vegetable greenhouse in Shouguang, Shangdong is not a rare example. PV technologies also have been applied to village lighting, pest killing and water-pumping systems. Let's take Xing'an County, Guiling City, in the Guangxi Autonomous Region of the Zhuang minority as an example. So far, 105 natural villages in the county have been endowed with 1,558 solar street lamps. This resulted in breakthroughs in solar insecticidal lights and solar pumps.

In conclusion, the PV revolution is the path to follow in order to improve production, our lifestyle, the sustainable utilization of rural infrastructures and environmental protection in rural areas.

The Sunshine Culture Reshapes the World

The second Industrial Revolution, represented by coal and oil, was based in a "black gold" culture. The third Industrial Revolution, inspired by PV, worships a sunshine culture.

In that sunshine culture, industries are developing from centralization to distribution, and organizations are shifting from being intensive to being flat. People will believe that resources are used but not owned by them, and they will advocate wealth-sharing rather than monopoly.

Solar PV will turn our mindset and actions upside down.

From Foes to Friends

The impact that PV exerts on culture and politics reminds me of a film, *Black Gold*. The film narrates a story which takes place in Arabia in the 1930s, as Americans find oil midway between two tribes' villages. The chiefs of the two tribes disagree with each other on whether to exploit the oil well. They have the following conversation:

The opponent (O): "The Quran does not allow us to use oil or cars!"

The proponent (P): "If Allāh forbids us to use oil, why does he allow oil to be buried under our feet?"

O: "What on earth do you what?"

P: "I want an oil-rich land."

O: "No way, I've promised it to. . ."

P: "You have no right to make promises; it does not belong to you."

O: "Is it yours?"

P: "It belongs to them (Arabian tribes)."

O: "Then what will you use it for?"

P: "I will build hospitals and schools."

The two tribes fail to reach a consensus and conflicts eventually escalate into a tragic war. The opponent and proponent represent the traditional Arabian culture and the "black gold" culture. The film is fiction. However, in the Gulf region, the evolution of the oil industry just might be even more intricate than the film.

At the beginning of the 1930s, after 20 years of war King Abdul-Aziz founded the Kingdom of Saudi Arabia, which was very poor at that point. A British advisor, Harry, told the king: "With numerous treasures lying under the ground, what is the point of being worried? You have abundant oil and gold underground, but you don't allow others to exploit them."

The king changed his mind and signed a contract with Standard Oil, agreeing to lease massive areas of land to the US company for 60 years. This agreement opened the Middle East to Western countries to exploit oil. And thus the "black gold" culture was transferred to the West with the sounds of tens of thousands of drills.

To the Arabians, "black gold" means avarice. In America, where the term was coined, "black gold" also has the same meaning. John Davison Rockefeller, an oil magnate who owned Standard Oil, was the forerunner who entered Saudi Arabia. He wrote a letter to his son: "If you carefully review history and human activities, you will get the conclusion that every society is established on greed."

I think Rockefeller's interpretation of greed is only valid in the second Industrial Revolution. Our generation is reading the last page of the second Industrial Revolution and turning to the first chapter of the third one. We will bid farewell to the "black gold culture" and welcome the "sunshine culture."

The sunshine culture is a culture of harmony. This means two things. The first is harmony between man and nature. "War against nature," "defeating nature," and "man vs. nature" are cultures that have been punished by nature. The broad application of solar power, "chlorophyll economy," and "green life" are the correct approaches for sustained development and the survival of mankind. The second meaning is harmony between people. This includes people in the same country, as well as people from different countries.

"We only have one Earth." This is the reason for friction when fossil energies rule the world. However, when we have infinite solar energy, when we have to deal with global warming together, when we share economic achievement, the same quote becomes the most practical reason for human beings to forget about conflicts and start thinking about cooperation. To start making friends instead of foes.

Reshaping the Scale of Economy

From foes to friends, we see the impact PV exerts on politics and culture. From an economic perspective, PV is going to reshape the scale of the world's economy.

In his book, *The Third Industrial Revolution*, Jeremy Rifkin summarizes world economic trends. He thinks that a flat world economy is the trend of the third Industrial Revolution. To be specific, the substitution of fossil energies by solar power will turn the focus of economies of scale from centralization to distribution. Centralized power generation will be replaced by distributed power generation; power consumers are turned into power producers as generation, supply and consumption of power will become the duty of the consumers themselves. There will no longer be a monopoly in the entire energy system, but coordination among multiple players in the game. The manufacturing industry will become decentralized, thus forever changing relations among manufacturers, suppliers, and consumers.

The Internet enables us to see such changes. Information delivery used to be an individual activity without any scale; later on, communication became a professional activity, whose professionalism was embodied in the centralized scale of communication. The marriage between the Internet and the economy gave rise to the "Internet economy," which promotes distributed scales. In 1999, Nike launched its online customization business. Every customer could choose different soles and colors from the options provided online and even print names up to 8 letters long on shoes. At the same time, online shopping platforms like Taobao came into being. According to statistics on Taobao, in the first half of 2012, the number of registered users totaled 470 million, and the online transactions in that year amounted to 800 billion.

Nike and Taobao are examples of two different industries justifying the forthcoming changes in the world economy. Judging from reality and trends, the PV revolution will transform the world economy mainly through the phasing out and reshaping of large companies.

We can draw a conclusion from previous Industrial Revolutions: energy revolutions trigger Industrial Revolutions which give rise to large companies. Many of the world *Fortune* 500 are energy companies or companies closely linked to the energy sector. Changes in energy structure lead to changes in economic models. It should be assumed that the PV revolution will transform those large companies.

Besides changes in the macro economy and large companies, the production model of human economic society will also be reinvented. We have witnessed modern industry replacing the handicraft industry. Industry, whose scale is based on uniformity, improving efficiency with rigorous organization, and making profits with scale and efficiency, defeated the handicraft industry, characterized by customized products, independent labor, and low efficiency.

The third Industrial Revolution will overturn this replacement. In essence, elements like the freedom of individual behavior and personality in the handicraft industry will be integrated into traditional industry in order to overcome the weakness of traditional industry. In a sense, the third Industrial Revolution will promote handicraft industry and customization and reject large industry and uniformity. The development of flexible manufacturing and the 3D printer has given us more possibilities in the future economy. In fact, mass production and customization contradict each other. The beauty of mass production is diminishing marginal cost. However, if we continue producing massively, we may suffer great cost from the customization of such products.

Reconciling the contradiction between mass production and customization depends on a flexible manufacturing system, which is composed of three sub-systems—information control systems, material logistics systems, and NC processing equipment. Based on group technology (GT), the systems aim at improving organizational management for multi-category and small-batch production, resulting in economic outcomes as good as those seen in mass production.

The fundamental principle of GT is that it is based on the similarities of the structural features of parts, the engineering process and processing methods. It systematically groups all product parts, putting parts with similarities into the same group, setting up different processing units targeting different parts with corresponding machine tools. Such re-assembly turns small-size batch productions into big-size ones, tremendously improving efficiency.

The BPIV revolution with flat reinvention will provoke a series of transformations in the world economy. Against this backdrop, old corporate systems and monopoly powers will be demolished, and large companies will have to promote more flexible production models, thus reshaping the scale and structure of the world economy.

From "Exclusive Ownership" to "Sharing"

The PV revolution will also change our mindset in regard to consumption. The Sunshine Culture will lead people to prefer "sharing" over "exclusive ownership." Things in limited quantity eventually trigger competition. As black gold, oil and coal evoke greed and desire, which can lead to pillage and war. However, pursuing something found in infinite amount is meaningless, and better utilization of such things becomes more important. Solar energy can never be exclusively owned by anyone.

In *The Third Industrial Revolution*, Jeremy Rifkin pays a compliment to car rental, online music services, and timeshare vacations, describing them as the "sharing" of the third Industrial Revolution and believing that they are changing our positioning and ideas about economic theories and time.

Rifkin thinks that, in the traditional capital market, profits are derived from margins of transaction costs. That is to say, in each link of the value chain, sellers charge customers more to reap profits. The end price of goods and services reflects this price hike. In the third Industrial Revolution, however, the transaction cost of information and energy is almost zero. It is impossible to make marginal profits. Therefore, we have to redefine profit.

Many others agree with Rifkin. For example, Rachel Botsman, in her book *What's Mine is Yours: How Collaborative Consumption Is Changing The Way We Live*, commented on the phenomenon of "3 million people in 235 countries seeking short-term accommodation via the website of Couchsurfng," saying "it has forced people to rethink private property, which, to some extent, might be a revolution as significant as the Industrial Revolutions."

However, I don't believe the "sharing" above is authentic "free sharing." For example, the free download of music and books mentioned by Rifkin is a misleading concept because it is a violation of intellectual property rights. Some major search engines allow people to upload resources, namely other people's works, to the Internet, making the search engine a huge database. Unauthorized uploads are considered copyright infringement. Would a new business model built on illegal activities be sustainable?

Chris Andersen, author of *The Long Tail: Why the Future of Business Is Selling Less of More*, wrote another book named *Free: the Future of a Radical Price*. He believes that a free business model on the Internet is emerging. He lists two free models: one is free for all, e.g., online news. Another is free for some users but charges other users money, thus providing them with better services and more information. As for the second model, (e.g., instant chatting software), software vendors first provide free services to users to gain popularity, and then develop value-added services and sell them to some of the users. The money paid by these users covers the operational cost and profits of the Internet company.

Based on the above analysis, even users of free services are assuming costs which are embedded in the sharing model, which is hard to identify because of the interest chain in the Internet business model. However, authentic sharing is not about gaining yields.

With changes brought to the economic system by solar power, exclusive ownership will become increasingly less important while sharing will become more significant. I'm not saying that there will be less exclusive ownership: sharing inherently ends with individual ownership. Sharing does not stand in the way of ownership, and people are free to select the information they need.

The new energy revolution combining the Internet and new energy together will create distributed power plants—relying on the omnipresence of solar power—and the smart interconnected grid, which will promote the culture of "sharing." In my opinion, sharing is not a business model, but a social ideal, the theoretical cornerstone of the third Industrial Revolution.

EPILOGUE

LIFE OF A PV PERSON

I would like to share a conclusion with you now after six chapters of expanding on my personal experience, observations, thoughts, and the data and cases I have collected.

Industrial Revolutions in history are driven by energy revolutions: the first Industrial Revolution was fueled by the substitution of coal for wood; the second was sparked by the replacement of coal by oil. And the third Industrial Revolution, which is around the corner, will be propelled by solar power, and that is the "ultimate substitution" in energy history.

Both Industrial Revolutions gave rise to new great powers. Both the British, rising in the first Industrial Revolution, and the United States, coming out of the second one, witnessed historic moments when coal replaced wood and oil substituted for coal. With the advent of the third Industrial Revolution, China is most likely to rise as a great power. One must now wonder: can China seize this opportunity and realize the "Chinese Dream"?

The answer is Yes. Competition in solar power is different from that in traditional energy. It is not a competition for resources, but a competition for core technologies. The one who controls core technologies controls energy. The kernel of the third Industrial Revolution is the new energy revolution and the crux of the new energy revolution is the application of solar power, whose nucleus is the PV revolution. The strategic direction of the PV revolution is thin film and flexibility, and China is leading the world in terms of PV technologies, PV companies, and national PV strategies. The "PV dream" is a powerful impetus behind the "Chinese Dream."

What will our life be like after the realization of the PV dream and the Chinese Dream? Next, let's envision the life of a "PV person."

A PV person is living in a solar house: a set of solar panels are installed on the rooftop and three sides of the house are covered with thin film. With big windows facing towards the sun, the occupants receive abundant but not dazzling sunshine in the room. The windows are made of thin film which can also generate power. To the inattentive eye, this would appear to be a pretty but normal house. As a matter of fact, solar power is integrated into the house in a subtle yet sophisticated way.

Opening the window, one can breathe fresh air and enjoy the blue sky, white clouds, twittering birds and fragrant flowers. It's a wonderful world without the pollution discharged by coal-fired power plants or the burning of fossil fuels. When strong wind blows, the PV person does not shut the window because mankind has left behind sand, smog and haze for good, all thanks to the improved environment.

All the electricity consumed by the appliances used in the house of the PV person is generated by the house itself. The person is no longer worried about the increased cost of modern appliances or electricity tariffs. However, at the beginning of each month, the PV person will still receive an electricity bill, which displays how much profit is made from the surplus electricity transmitted back to the national grid. What a wonderful world! The profit gained from surplus electricity generation may not be substantial. However, they are a welcome compensation to the household.

It's worth noting that, for 9-to-5 PV persons, getting up early to work or getting trapped in traffic jams is not a necessity anymore. They work at home.

Imagine the telephone ringing suddenly. The PV person answers the phone and it turns out to be a friend asking him out. Then he goes to the garage, chooses a car and checks the battery of the vehicle. Satisfied with the battery that he can travel 100 km on, the PV person drives away. A PV car is also a power generation system by definition which can recharge anytime on sunlight while moving on the road.

He wants to call his friend on the way, but the phone battery is low. He then uses a solar power charger to recharge his cell phone.

Meeting each other, they decide to take a long trip. They arrive at a special airport on time, and directly fly to Sanya on a solar aircraft. Three days later, they travel to Guangzhou on a solar train and return to Beijing where the PV person lives.

After the trip, they decide to have another adventure—climbing Mt Qomolangma. However, they forget to bring oxygen tanks with them. No big deal, they decompose mountain snow with a solar power generation tool to produce oxygen.

When the PV person lives in the context of the third Industrial Revolution, his personal life will be different and probably described as follows.

8:30 AM, Silicon Valley, Nell Glenflore was busy with sharing. He asked a babySitter who worked for both Nell and his neighbor to take care his 15-month-old son. Mountain View, where Nell lives, is only several blocks away from the California Railway Station. In a café, he logs onto the website of Lending Club, and give out some small loans. A loanee needed money for his wedding; another one planned to start a pet shop; a third one was about to relocate. Riding his bicycle to the railway station, he gets on an electric Prius, which he reserved for several hours from a shared car company. He drives to Berkeley University, and after visiting a residential co-op, he comes to a shared office. He works here once a week. "When people begin sharing, they start to think what to share next." Nell says. "Small changes eventually lead to great transformation."

This is the opening paragraph of an article titled *Age of Sharing* published in *Readers*. Almost all the comments on the third Industrial Revolution point out this new business model rooted in a new culture. Mark Levine, a journalist at the *New York Times*, believes that "the relation between sharing and ownership is like the one between iPod and cassette, and the one between solar cell and coal. Sharing is clean, new, urbanized, and post-modern, while exclusive ownership is sluggish, selfish, timid, and backward."

Today, with PV technologies, such a wonderful life is not far away. As long as we have dreams, we can live such lives.

All in all, in the 21st century, we can fulfill our dreams with sunlight in the context of the third Industrial Revolution and a rising China.

The 18th CPC National Congress depicted the Chinese Dream. The PV revolution is both an important component and a cornerstone of the Chinese Dream. President Xi Jinping said that we have the dream and opportunity, so we should strive for the dream. The PV revolution is our dream, our opportunity and the goal we strive for.

China will certainly have a brighter future. Only when we materialize a PV China, can we realize the dream of scientific development and peaceful rise, and make China a civilized great power leading the world in this new sunshine culture.

NOTES

Preface

1. 1 USD equals 6.05 yuan (exchange rate in January 2014).
2. 1 GW=1000MW=1×10⁶kW.

Chapter One

1. 1Euro equals to about 8.21 yuan (exchange rate in January 2014).
2. *When Will Oil in the World be Exhausted*, *Lianhe Zaobao*, October 25, 2009.

Chapter Two

1. *Energy and Environmental Challenges and Solutions of China in 2012*, Wang Jinnan, Chief Engineer of the Chinese Academy for Environmental Planning (CAEP), and the taskforce of CAEP, *Economic Information Daily*, November 2005.

Chapter Three

1. *Shale Gas Development: an American Story*, Wu Yun, the *People's Daily*, April 2, 2013
2. A Ponzi scheme is a fraudulent investment operation where the operator pays returns to its investors from new capital paid to the operators by new investors, rather than from profit earned by the operator. The Ponzi scheme is the oldest and most common investment fraud and is a variant of the pyramid scheme.
3. *China Should be Calm with the Shale Gas Revolution*, Zhong Shi, *China Youth Daily*, January 7, 2013.
4. *Powering the Future: A Scientist's Guide to Energy Independence*, Daniel B. Botkin and Dana Perez, translated by Cao Mu, Publishing House of the Electronics Industry, 2012.
5. *Earth: The Sequel: The Race to Reinvent Energy and Stop Global Warming*, Fred Krupp and Miriam Horn, translated by Chen Maoyun, Oriental Publishing House, 2010.

6. *The PV Ambition of Japan*, http://solar.ofweek.com/2012-09/ART-260006-8500-28641131.html

7. *Lessons Learned from the PV Market in Japan*, http://www.chinairn.com/news/20121212/468105.html

8. Usually known as TÜVs, short for the German Technischer Überwachungs-Verein (Technical Inspection Association)—German organizations that work to validate the safety of products of all kinds to protect humans and the environment against hazards.

9. *Analysis of the PV Industry in Korea, 2013, Photovoltaic Energy Industry Observer.*

10. *Technology and Industrial Chain: Two Dilemmas of the PV Industry in Japan*, China Semiconductor Industry Association.

11. *Sino-EU PV Price Commitment Kicks Off Today*, Guo Liqin, *China Business News*, August 6, 2013.

12. *EU Lost the PV Trade War, Wall Street Journal.*

13. Used as the central processing unit (CPU) of many workstations and high-end personal computers of the time.

14. The seven strategic sectors include energy-saving and environmental protection; next generation information technology; biotechnology; advanced equipment manufacturing; new energy; new materials; and new-energy vehicles.

Chapter Four

1. *Economic and Social Achievement of the New China*, www.gov.cn

2. Purchasing Managers' Indexes (PMI) are comprehensive economic indicators derived from monthly surveys of private sector companies, including manufacturing PMI, service PMI, and construction PMI. PMI reflects changes in economic trends. PMI>50 means economic development; PMI<50 means economic downturn.

3. "Quotes from a research report of the Solar System Institute of Fraunhofer-Gesellschaft and a financial report of First Solar, Inc."

Chapter Five

1. *The Lost LDK, 21st Century Business Review*, August 2012.

2. *Collapse of LDK Will Destroy 10 Years' Economic Growth in Xinyu, China Business Journal*, January 2013.

3. *The Industrial Chain Dilemma Faced by PV Industry*, the *Energy Magazine*, October 2010.

4. *PV Tycoons Meet in Shanghai to Fight Against US Anti-dumping and Anti-subsidies Investigations, 21st Century Business Herald*, May 2012.

5. Same as 4.

6. *Access Problem of PV Power Stations, Energy*, March 2013.

7. *State Grid Bails out PV Companies, Innovative Finance Observation*, October 2012.

8. *Comparing the Advantages between Thin-film Cells and Polycrystalline Silicon, China Industrial Economy News*, September 2011.

9. Same as 1.

10. *From the Richest to Broke, Money Week*, March 2013.

11. *The Resurrection of Suntech*, the *Economic Daily*, August 2013.

12. *What Does the Development of German PV Industry Tell Us*, published on the website of the Ministry of Commerce, February 2012.

13. *PV Policies Are Made Based on Wisdom from Successful Experience*, the *China Energy News*, November 2012.

For Reference

1. *Powering the Future: A Scientist's Guide to Energy Independence*, Daniel B. Botkin and Dana Perez, translated by Cao Mu, Publishing House of Electronics Industry, 2012.

2. *Energy Revolution Changes the 21st Century*, Liu Hanyuan and Liu Jiansheng, China Yanshi Press, 2010.

3. *Study on New Energy Industry in China*, Zhang Qin and Zhou Dequn, Science Press, 2013.

4. *Analysis of Power Generation by New Energies, 2012*, the National Energy Institute of the State Grid, China Electric Power Press, 2012.

5. *New Development of Energy Laws of Different Countries*, Hu Xiaohong, Xiamen University Press, 2012.

6. *Earth: The Sequel: The Race to Reinvent Energy and Stop Global Warming*, Fred Krupp and Miriam Horn, translated by Chen Maoyun, Oriental Publishing House, 2010.

7. *Who Won the Oil Wars*, Andy Stern, China Citic Press, 2010.

8. *Great Powers and Industrial Wars*, Yao Xiaohong, Xinhua Publishing House, 2012.

9. *Decode New Energy*, Zhang Shuai, Xing Zhigang, and Yao Yao, Wenhui Press, 2011.

10. *Why Your World Is About to Get a Whole Lot Smaller: Oil and the End of Globalization*, Jeff Rubin, translated by Zeng Xianming, China Citic Press, 2011.

11. *Green Capital—Analysis of New Energy Industries in China*, Bo Feng and Chi Xiaohong, Posts&Telecom Press, 2013.
12. *Shale Gas Development*, Kaitlyn M Nash, translated by Wang Lihua and Zhou Jing, Shanghai Science Technology Press, 2013.
13. *Energy Development Roadmap of China by 2050*, the Energy Strategic Team of CAS, Science Press, 2009.
14. *Thin-film Cells and PV Power Stations*, Duan Guangfu and Duanlun, China Machine Press, 2013.
15. *The Third Industrial Revolution,* Jeremy Rifkin, translated by Zhang Tiwei and Sun Yuning, China Citic Press, 2012.
16. *China Should be Calm with the Shale Gas Revolution*, Zhong Shi, the *China Youth Daily*, January 7, 2013.
17. *US PV Industry Blocked by Trade Protectionism*, Wang Zongkai and Yang Jian, Xinhua Net, 2012.
18. *Analysis on the On-grid Tariff Model of Germany*, Cao Shiya, *China Electric Power News*, March 7, 2013.
19. *Comprehensive Analysis on the PV Market in Korea, 2013*, anonymous, *PV Industrial Observer*, April 22, 2013.
20. *Advantages of the PV Companies in Japan*, consultant of China Investment Corporation, Phoenix Net, December 6, 2012.
21. *Technology and Industrial Chain: Two Dilemma of the PV Industry in Japan*, China Semiconductor Industry Association, April 18, 2012.
22. *PV Industry Development Report*, Li Tianyu, September 1, 2011.
23. *China PV Industry Report 2013*, SEMI PV Advisory Board, March 27, 2013.
24. *Development Report on China and Overseas PV Industry*, 2011, anonymous, November 1, 2011.
25. *Global New Energy Report*, Hanergy Group, June, 2013.
26. *Accelerated Consolidation of PV Industry in China*, Zero Power Intelligence, July 31, 2013.
27. *EU Lost the PV Trade War, Wall Street Journal*, August 2, 2013.
28. *Research on Thin-film Cells*, Chen Gangxiang, May 13, 2012.
29. *Sino-EU PV Price Commitment Kicks Off Today*, Guo Liqin, *China Business News*, August 6, 2013.
30. *The Overestimated Shale Gas*, anonymous, *Hong Kong Economic Daily*, August 5, 2012.

INDEX

273

ABOUT THE BOOK

This book discusses how China can tackle its energy bottleneck and achieve a sustainable, peaceful rise through a photovoltaics (PV) revolution.

In 2012, China's dependency on foreign crude oil was more than 50%. Its energy security and independence has become a cause for concern. As the country is increasingly connected with the rest of the world with more complicated international relations than ever, instability and risks are also mounting with a rising possibility of conflicts.

At the beginning of 2013, China was hit hard by a dense wave of smog. In Beijing and many other northern cities, the level of particulate matters in the air with diameters of 2.5 micrometers or less reached hazardous range. The country faces unprecedented environmental pressures today. Solving this problem not only concerns people's well-being, but also China's responsibility as a major nation with an important role in the world. Climate negotiations are not only about reducing carbon emissions, they also involve each country's development strategy and economic benefits.

China is confronted with an energy and environment challenge not seen previously in its history. However, a challenge often brings with it an opportunity.

In human history, industrial revolutions are often initiated by a combination of new energy sources and communications technologies and followed by a reorganization of the global power structure. The UK and the US seized the opportunity of the first two Industrial Revolutions in history and became the world hegemonies.

Today, new energy sources represented by PV combined with information technology are giving birth to a third Industrial Revolution, presenting a new opportunity to change the global power structure. In this new industrial revolution, China, for the first time, is leading the race. And this country is absolutely capable of getting ahead.

In this book, the author puts forward the PV revolution as an innovative means for China to tackle its energy bottleneck, transform its economy, and grow sustainably as a world leader.

The author first provides a summary of trends in new energy revolutions worldwide, including lessons learned from Europe, the United States, Japan, and South Korea. Then he looks into the tortuous path of development in China's PV industry during the recent decades since China's reform and opening up. Incorporating theories, practices, and mindsets from the vantage points of economy, society, industry, and business, the author discusses current and future strategies, policies, and measures that China should take in this round of industrial revolution. Finally, he makes the prediction that China will lead the world in the PV industry.

National revival has been a dream of the Chinese people since modern times. We believe that PV will contribute to a more beautiful China by fundamentally changing how we perceive and utilize energy.

The sun shines all over the world. We are using sunshine to realize our dream.

ABOUT THE AUTHOR

Li Hejun was born in 1967 into a Hakka family in Heyuan City, Guangdong Province. He is now the Vice President of All-China Federation of Industry and Commerce, member of the National CPPCC and the President of China New Energy Chamber of Commerce. He is the Chairman and CEO of Hanergy Holdings Group.

As a renewable energy entrepreneur, Li launched the "Global Replacement Initiative" for clean energy to replace fossil fuels. Both a new energy strategist and practitioner, he is among the first to advocate for distributed generation in China and to popularize the idea of consumers' generating electricity for their own needs and sending the surplus back into the power grid.

Li founded Hanergy in 1994 with an ambition to give back to the country by developing its industrial sector. With a firm belief that clean energy can change the world, he led Hanergy to complete the Jin'anqiao Hydropower Station, the world's largest hydropower station invested in and built by a private company. In 2006, he made his first move into the photovoltaics (PV) industry and strategically predicted that thin films and flexibility are the future trends of the PV industry. Today, Hanergy has grown into a global leader in thin-film solar panel manufacturing, both in terms of output and technology.

"In 2035, clean energy, as represented by solar power, will replace 50% of fossil fuels." Based on this prediction, Li suggests that PV will become a pillar among new energy sources, play an important role in the third Industrial Revolution, and profoundly change the way we live and work. In this process, China has every reason to move ahead of others. PV will become an engine for the rise of China and for the realization of the Chinese dream.

Solibro Industrial Park, Germany

Solibro is a pioneer in the commercialization of copper indium gallium selenide (CIGS) cell technology. The average conversion rate for component production is 14.7%; the highest recorded rate was 18.7%. Solibro is a top-level international enterprise, and one of the first to produce CIGS cells commercially. It is completely UL- and IEC-approved.

Thin-film PV installation made of Solibro components

Integrated PV building located in Uppsala, Sweden, using thin-film solar components produced by Solibro

Exhibition booth in an English IKEA for Hanergy's small domestic solar power systems

In England, for the sum of £4950, a family can install 25 thin-film PV panels (a 3kW system) and on average produce 3,600kWh of electricity every year. This kind of investment has three benefits: (1) it allows the family to receive £576 annually in government subsidies; (2) it allows the family to save £261 in electricity annually (half of the system's capacity is for private usage); (3) the income from selling electricity is £81 (excess power can be sold at the price of £0.045 per kWh to power companies). In general, a family can make a profit of £918 annually, for an annual return on investment of 13%.

Workers install Solibro thin-film PV components on a residential roof in England.

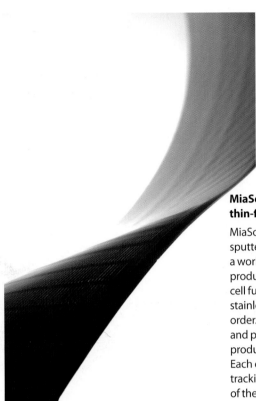

MiaSole copper indium gallium selenide (CIGS) thin-film components

MiaSole's unique unbreakable vacuum continuous sputtering technology is what made its thin-film cells a world leader. MiaSole uses a flexible roll-to-roll production technique which allows them to deposit all cell function membranes on a 50-micron thick flexible stainless steel substrate, and then cut and select to order. Interconnected cells guarantee high reliability and performance. A precise control over the industrial production process assures a high yield and lost cost. Each cell has a unique identification code, which allows tracking during production and utilization. The first part of the production process of the whole battery takes only 40 minutes. MiaSole employs an agile technology which allows cells to be packaged as flexible or rigid components. These highly efficient components have been UL- and IEC-approved.

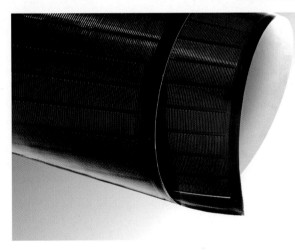

Miasole's CIGS thin-film components (US)

Miasole's flexible thin-film components can be widely used on agricultural greenhouses in vast rural regions. It is the key to upgrading the rural power grid and solving the issue of electricity-less populations in remote regions.

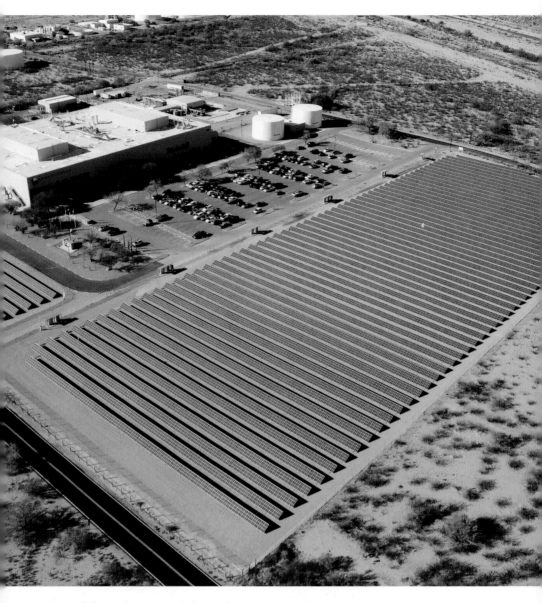

GSE (Global Solar Energy) is located in Tucson, Arizona, USA.

 It launched its solar energy pilot project in 2003, building power stations on its factory's land. GSE has always been a leader in the development of portable flexible solar components, and possesses the individual intellectual property rights for its CIGS inhomogeneous coating technology. GSE's flexible components have the highest conversion rate in the world at about 12.5%, with a highest recorded rate of 15.8%. The enterprise plans to increase its component efficiency to 17% in the next few years.

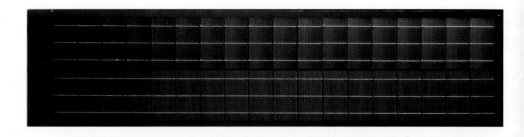

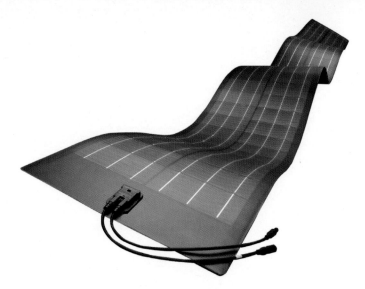

GSE flexible components

GSE produces lightweight components which do not damage the overall structure of roofs when installed and do not require the installation of expensive supports. They can be widely used on all types of buildings around the world, including light steel roofs which must not be damaged or do not allow structure installation, large flat roofs with weight limitations, as well as undulating roofs which do not allow for conventional rigid components. It can even be installed on car surfaces.

Japan's rooftop power stations

This power station uses GSE's flexible thin-film solar components and weighs 3 kg per square meter.

Yokosuka project, Japan

These flexible thin-film solar cells are soft and bendable, and can be used on all types of curved surfaces.

GSE's flexible thin-film solar component PowerFLE™ being used on an RV's roof in Arizona, USA.

Residential rooftop power stations in Michigan and Colorado, USA

The solar components used are Dow Solar's PV shingles for which GSE produces chips. The component is light and can be used on slanted roofs.

Hanergy's PV wall at its PV base in Wujin, Jiangsu.

On May 10, 2012, Hanergy's wall was successfully connected to the grid with an installed capacity of 66.3 kW.

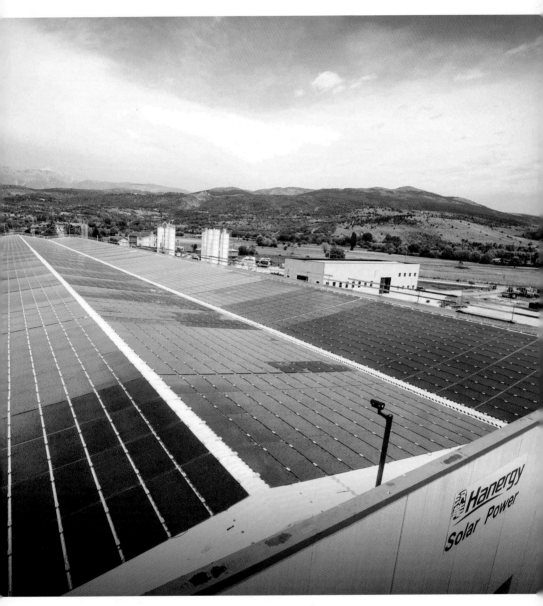

Rooftop power stations in L'Aquila, Italy.

L'Aquila's biggest rooftop solar power station is also a symbolic part of Italy's "Harmful Asbestos Tiles Removal Campaign."

Hanergy's PV power station in Gonghe County, Hainan Tibetan autonomous prefecture

The power station is located in the PV industrial park of the Qabqa village of Gonghe county in the Hainan Tibetan autonomous prefecture (Qinghai province). It is currently the largest thin-film PV surface power station in the world.

Hanergy's "Golden Sun" pilot project in Heyuan, Guangdong

Total installed capacity equals 10 mW. The project has been connected to the grid since April 12, 2013.